Label Dispensing and Application Technology

Other Labels & Labeling books:

ENCYCLOPEDIA OF LABEL TECHNOLOGY
Michael Fairley

THE HISTORY OF LABELS
Michael Fairley and Tony White

DIGITAL LABEL AND PACKAGE PRINTING
Michael Fairley

ENVIRONMENTAL PERFORMANCE AND SUSTAINABLE LABELING
Michael Fairley and Danielle Jerschefske

CONVENTIONAL LABEL PRINTING PROCESSES
John Morton and Robert Shimmin

LABEL DESIGN AND ORIGINATION
John Morton and Robert Shimmin

LABEL DISPENSING AND APPLICATION TECHNOLOGY
Michael Fairley

CODES AND CODING TECHNOLOGY
Michael Fairley

LABEL EMBELLISHMENTS AND SPECIAL APPLICATIONS
John Morton and Robert Shimmin

BRAND PROTECTION, SECURITY LABELING AND PACKAGING
Jeremy Plimmer

DIE-CUTTING AND TOOLING
Michael Fairley

MANAGEMENT INFORMATION SYSTEMS AND WORKFLOW AUTOMATION
Michael Fairley

SHRINK SLEEVE TECHNOLOGY
Michael Fairley and Séamus Lafferty

LABEL MARKETS AND APPLICATIONS
John Penhallow

For the latest list please visit: **www.labelsandlabeling.com**

Label Dispensing and Application Technology

Michael Fairley LCG, FIP3 and FIOM3

Label Dispensing and Application Technology

First edition published 2015 by:
Tarsus Exhibitions & Publishing Ltd

Printed by CreateSpace, an Amazon.com company.

ISBN 978-1-910507-05-6

Contents

While every care has been taken to ensure the information, charts, diagrams and illustrations in this publication are correct at the time of publishing it is possible that technology, specifications, markets and applications, or terminology may change at any time, or that the author's or contributor's research or interpretation may not be regarded as the latest accepted guidance in some parts of the world of labels.

The publishers therefore cannot accept responsibility for any errors of interpretation or for any actions, decisions or practices that readers may take based on the publication content and would advise that the latest industry supplier specifications, standards, legislative requirements, performance guidelines, practices and methodology should always be sought before any investment or implementation is made.

Preface

Pressure-sensitive labels have experienced a quite dramatic growth over the past 30 to 40 years. From just a few percent market share of all label technologies in the late 1970s, pressure-sensitives have come to dominate the label market in the developed world, and even in emerging markets now make-up one third or more of all label usage.

There are many reasons for this growth: the range of pressure-sensitive materials available; the diversity of printing and finishing technologies; simplicity and ease of use; production line flexibility; the range of sizes and shapes that can be labeled; the ability to overprint and code at the last minute; and a competitive applied label cost.

Much has been written about the extensive range of pressure-sensitive substrates, and all the conventional and digital label printing, finishing and die-cutting processes that can be used. But there are few if any detailed analyzes of label dispensing and application technology.

Yet without modern label dispensers, hand labelers, print and apply solutions, semi-automatic, fully automatic and rotary label applicators and linerless application systems, there would not be a pressure-sensitive label industry. They are what makes pressure-sensitive labels so attractive to the food, healthcare, pharmaceutical, wines and spirits, chemical, industrial and other end-usage sectors.

This handbook aims to explain and de-mystify label dispensing and application technology. How is it possible to separate pressure-sensitive face material from the silicone backing? How is label application controlled? How can labels be placed around corners or into recesses? How are products moved into the correct position for labeling, held steady or rotated on the label application line? How can last-minute price, weight, graphics or codes be added? All these questions are answered.

A final chapter sets out to look at troubleshooting on the applicator or application line, reviews where things may go wrong and why, and provides some basic problem solving guidelines. Commonly, it is not the applicator that is at fault. More likely the problem or fault will be related to the choice of substrate or to faulty or poor die-cutting, rewinding or handling. All these factors are important for material suppliers, converters and end-users to understand.

If the handbook leads to fewer application problems and more satisfied label customers, then it will have justified the time and effort put into writing it.

Michael Fairley
Director, Labels & Labelling Consultancy
Founder, Label Academy

About the Label Academy

This book is part of the recommended study material for the Label Academy, a global training and certification program for the label industry. The Label Academy was created by the team behind Labels & Labeling magazine and the Labelexpo series of events.

The Academy consists of a series of self-study modules, combining free access to relevant articles and videos with paid text books (both printed and electronic). Once a student has completed a module, there is an opportunity to take an online test and earn a certificate.

It is expected that a Label Academy qualification will become a standard in the industry – for printers/converters, suppliers, brand owners and designers – and assist in providing a benchmark. In addition to its own training, the Label Academy will aim to become a resource provider to the many existing educational programs in the industry. Accredited training courses will be promoted through the Label Academy website and books will be provided at discounted rates.

The Label Academy concept was pioneered by industry expert Mike Fairley. This was in response to a reduction in the number of dedicated printing colleges and the need to standardize training across the world. The label industry also has its own specific training needs – it has some of the widest range of materials, printing processes and finishing solutions of any printing sector.

We are also working with other training experts and authors to ensure that the Label Academy provides up-to-date and relevant training material for the industry.

The Label Academy is supported by the key trade associations, including FINAT, TLMI and the LMAI.

www.label-academy.com

Label Academy sponsors

Thank you to our founding sponsors, without whom this ambitious project would not have been possible:

Cerm

Cerm designs business automation software solutions to meet the specific demands of flexo and digital narrow web printers. Using the latest technology, our team's focus is on innovation and continuous improvement.

Our automation solutions support each step in the printer's integrated workflow – from estimating to production, shipment and data collection – and provide the feature and functionality printers need to gain efficiency and improve profitability.

Cerm inspires collaboration and helps printers remain competitive in the market and deliver the best products possible. We are proud to sponsor the Label Academy and contribute to the future of the narrow web printing industry.

www.cerm.net

Flint Group Narrow Web

Flint Group Narrow Web has the products, the solutions, and the technical experts to handle any print situation. Providing solutions for food packaging, sustainability, increased bottom line, efficiency, and uptime – delivering the basics needed to run a successful operation, and the expertise to go above and beyond to another level of success.

Our experts provide solutions to your printing problems with the innovative products and services that have made us an industry leader around the world. Wherever you are, we are – available to help you reach your business goals today and into the future.

Continuous improvement is paramount to Flint Group; we are proud to sponsor the Label Academy and the benefits it will bring to the future of our industry.

www.flintgrp.com

Gallus Group

The Gallus Group with its production sites in Switzerland and Germany is a leader in the development, production and sale of narrow-web, reel-fed presses designed for label manufacturers. The machine portfolio is augmented by a broad range of screen printing plates (Gallus Screeny), globally decentralized service operations, and a broad offering of printing accessories and replacement parts. The comprehensive portfolio also includes consulting services provided by label experts in all relevant printing and process engineering tasks. The Gallus Group is a member of the Heidelberg Group and employs around 430 people, of whom 253 are based in Switzerland. The group headquarters is in St.Gallen, Switzerland.

www.gallus-group.com

PRINTING PRODUCTIVITY

MPS Systems B.V.

Producing high-quality label printing depends on several factors; one of them is the operator of the press.

As a press machine builder since 1996, MPS Systems B.V. knows how important training and education on subjects like pre-press, label printing and finishing is. For label printers, it is critical that their operators keep up with pre-press and press developments in addition to label trends. Therefore, MPS sponsors the Label Academy, to advance operator's passion for printing, share expertise and help multiply benefits.

The MPS slogans of 'Printers First' and 'Technology with Respect' have always underlined the core philosophy of MPS from press design to operator satisfaction. We develop our presses with a strong focus on user-friendliness and respect for the press operator: Printers First.

www.mps4u.com

HP Indigo

HP Indigo is a global leader in digital printing, with a broad portfolio of digital presses and workflow solutions. Indigo's proprietary Liquid Electrophotography (LEP) technology delivers exceptional print quality for the widest variety of applications including labels, flexible packaging, shrink sleeves and folding cartons. HP Indigo's digital presses match gravure print quality satisfying the most demanding brands.

A division of HP Inc.'s Graphics Solutions Business, Indigo serves customers in more than 122 countries, including many of the top label and packaging converters worldwide.

www.hp.com/go/labelsandpackaging

UPM RAFLATAC

UPM
The Biofore
Company

UPM Raflatac

In a little more than three decades, UPM Raflatac has become one of the world's leading manufacturers of pressure sensitive label materials, developing and leveraging the latest innovations in adhesive technology. Our film and paper label stocks are used for product and information labeling across a wide range of end-uses – from pharmaceuticals and security to food and beverage applications.

We are an engineering driven company with industry-leading products known for their consistent high quality and top performance. We are also known for the high performing supply chain and undisputed leadership in the area of sustainability. UPM Raflatac's dedication to innovation, sustainability and top quality is matched only by our commitment to service excellence. We call it the Raflatouch.

www.upmraflatac.com

About the author

Michael Fairley
Director, Labels & Labelling Consultancy
Founder, Label Academy

Michael Fairley has been writing and speaking about label and packaging materials, technology and applications since the 1970s, both as the founder of Labels & Labeling and other print industry magazine titles and as an international consultant writing or contributing to label industry market and technology research reports for the likes of Frost & Sullivan, Economist Intelligence Unit, Pira, InfoTrends and Labels & Labelling Consultancy.

He is the author of the Encyclopedia of Label Technology, co-author of the Encyclopedia of Brand Protection, a contributing author to the Encylopedia of Packaging Technology and a contributing author to the Encylopedia of Occupational Health and Safety. He also provided significant input to the Academic American Encylopedia.

He now works as a consultant to Tarsus Exhibitions & Publishing – which organizes the Labelexpo shows, Label Summits and publishes Labels & Labeling magazine – as well as regularly speaking at industry conferences and seminars.

He is a Fellow of the Institute of Packaging / Packaging Society, Fellow of IP3 (formerly the Institute of Printing), a Freeman of the Worshipful Company of Stationers, an Honorary Life Member of FINAT and a Licentiate of the City & Guilds of London Institute. He was awarded the R. Stanton Avery Lifetime Achievement Award in 2009.

Acknowledgements

It was back in 1986 that Labels & Labelling Consultancy first researched and wrote an analysis of self-adhesive label applicator technology. Yet the basics of label dispensing, sensing, web and product control are still largely unchanged. This original report therefore provides the starting point for this handbook.

Considerable research and analysis was undertaken to bring the handbook up-to-date, with extensive study of application equipment websites, supplier literature, magazine articles and industry-wide presentations. Contact was also made with many companies to request suitable technical or illustrative material.

The author wishes to acknowledge the valuable support, encouragement and assistance provided by these companies and organizations and, in particular, would like to thank the following for providing illustrative material: Accraply, Catchpoint, ETI Converting, Herma, ILTI, Ritrama, Start International and Universal Labeling Systems. Thank you also to Greg Smye-Rumsby of Concept Design and James Wenman of Tarsus who created most of the technical diagrams.

Special acknowledgement must also be made to Mike Cooper of Catchpoint for reading through all the draft chapters and providing feedback and additional information to ensure that the book contents reflect current technology and practices.

Chapter 1

—

Introducing label dispensing and application technology

—

The rapid growth in the use of pressure-sensitive labels for product decoration, marking and identification has been one of the success stories of the last fifty years. Also known today as self-adhesive labels, they offer labeling flexibility, simplicity of application without the use of wet adhesives, quick label applicator line changeovers and the potential to add coding, marking and overprinting close to the point of application on almost any size or shape of product.

—

Continuous innovation and marketing by substrate suppliers, technology manufacturers and converters has done much to make pressure-sensitive labels so widely used. Equally important was the development of technologies for die-cutting labels to shape, removing the matrix waste and then being able to accurately dispense and apply individual pressure-sensitive labels to a wide variety of product shapes and sizes. This was all pioneered by R. Stanton Avery, who reportedly used a matchbox as a peel plate or beak. Today's global industry is founded on these same principles.

LABEL DISPENSING MACHINES

From simply peeling the backing away by hand have come dispensing machines that will do the same job by pulling the label forward, either mechanically or electrically, moving the liner so the label to then be applied by hand. An example of this type of dispensing machine can be seen in Figure 1.1.

Hand dispensers like this have an unwind spool

Figure 1.1 - Hand dispenser. Illustration courtesy of Start International

that feeds the label web to a stripping plate, dispensing the label for hand application, with the backing liner taken away to be re-wound, disposed of or recycled.

Hand dispensers will normally be on a packing table or bench, with the product to be labeled taken to the machine for the label to be applied. With larger items the dispenser may be mobile and taken or wheeled to the product.

HAND-HELD LABELERS

From the early, relatively simple machines used to dispense and apply labels by hand, have evolved a variety of hand-held labelers – including those that can print one, two, or three lines of fixed, variable or sequential information before dispensing. Hand labelers, including those with print capabilities, need to be both robust enough to withstand continuous handling, and relatively lightweight to avoid hand and arm strain. An example of a modern hand labeler can be seen in Figure 1.2.

Figure 1.2 - A manual hand-held label applicator/labeler courtesy of Start International

Like hand dispensers, hand labelers will incorporate a label unwind spool, a stripping plate to remove the backing and feed the label forward for application, and usually, a backing liner spool to take up the

waste. Some early models used sprocket holes in the webs to allow the liner to be ratcheted forward to create the pull around the beak or peel plate. Hand labelers are designed to be portable and taken to the product.

Hand labelers, and those with simple one, two or three line print capabilities, come in a variety of designs, shapes and widths, but are limited in the length and width of labels that can be applied by the need for portability and ease of handling.

Semi-automated or fully-automated labeling machines are used to apply pressure-sensitive labels from a roll of labels onto a container or product, supplemented by the latest generations of high-speed 'intelligent' and rotary applicators and sophisticated labeling lines.

AUTOMATED AND 'INTELLIGENT' APPLICATORS

These more advanced machines may be controlled by touch-screen interfaces, microprocessors and use programable logic controllers, servo drives, sophisticated label or product sensing devices – both contact and non-contact – scanners and vision inspection systems, as well as having the potential to add a whole variety of mechanical or electronic overprinting devices into the labeling head or labeling line.

All types of pressure-sensitive dispenser or applicator have a number of key things in common.

Firstly, a means of peeling the release liner or backing away from the label. This is usually accomplished by unwinding a reel of die-cut, or sometimes butt-cut, labels and then pulling the web under tension around a peel plate or beak. As the backing is taken around the sharp angle of the plate, the front edge of the label is 'released'. Advances in high-speed release have been vital in developing the latest generation of high-speed applicators.

Once the label starts to be released from the backing, there are various ways of feeding it forward and pressing (applying) it to a pack, container or product. A number of different devices – including rollers, brushes, tamp pads or air jets – are used to ensure the label sticks with a positive contact to the product or container.

Figure 1.3 - An Accraply label applicator that can be integrated into existing packaging lines

Figure 1.4 - Herma 362M labeling system used to label Meira's redesigned herbs, spices and dressings

Figure 1.5 - Labels for Finnish food producer Meira are centerd exactly in a molded container recess in the tapered bottles

The effectiveness of the labeling operation is determined by adhesive tack levels and by the uniform application of pressure to the label, which ensures a positive contact of the adhesive with the surface of whatever item is being labeled.

Pressure-sensitive labeling lines may also be used in conjunction with fixed or variable speed conveyors, product handling devices, notch detectors, jigs, belts and plungers to take products into and from the label application heads, turn them, orientate them, place them in a recess, apply labels in particular places or angles, go around corners and take labeled products onto other finishing or packing operations.

In the labeling system shown in Figure 1.4. for example, the label application was a challenge due to a tapered label shape and size that could easily give rise to inconsistencies during the dispensing operation.

The labels had to be applied exactly into a molded container recess on a range of tapered bottles, which feature an oval cross section (Figure 1.5).

A special centering device on the applicator was used to position the label on the fly, with the product being arrested and centered during the labeling operation, so creating an unusual and interesting look.

Methods of applying a label to a product or container have increased in sophistication and variety, using rollers and brushes, pressure pads, air jets and pneumatic plungers to label almost any shape of bottle, jar, made-up or flat-pack carton, sandwich pack, tube or flexible packaging, right up to large boxes, drums, containers and pallets.

Using multiple labeling heads it is possible to apply two, three or more labels to a product or pack in one pass, with a typical placement accuracy of +/- 0.5 to 1.00 mm.

Applicators may also include automatic switching when a reel is spent, or when an applicator develops a fault, so the line does not have to stop. Automatic reel splice systems provide continuous running.

RANGE OF MODELS AND MANUFACTURERS

Application machinery manufacturers often specialize in equipment for specific markets or applications. These include pharmaceuticals, cosmetics and toiletries, compact discs, videos, weigh-price labeling of fresh foods, shipping address labels, drums, pallets, and labeling of newspapers and magazines. Labels have to operate in a variety of environments including heavy duty applications, low or high temperature and hazardous or 'flame-proof' environments. Tamper-evident shapes or security seals may be required in pharma, food or medical applications. Examples of some of the key pressure-sensitive tamper-evident labeling solutions can be seen in Figure 1.6.

Figure 1.7 - Illustration shows a typical pricing hand gun labeler

Figure 1.6 - Examples of tamper-evident labeling solutions

Labeling machines are also designed for personal or home use, for producing name badges or for use in filing systems.

OVERPRINTING AND CODING

Applicator lines now offer the possibility to add price, 'sell by', barcode, time or date stamps, product description and tracking information using a hand gun labeler (Figure 1.7) or print head on the application line. Techniques as diverse as hot-foil, cold foil, thermal, ink pad, inkjet, laser, and even flexo or letterpress are in use.

Other labeling lines can apply tamper-evident labels across the neck and cap of a bottle at the same time as applying front and back labels, or can add a

hologram label or on-sert a sophisticated booklet/leaflet label.

WIDE VARIETY OF SOLUTIONS

A wide variety of label application solutions are available, including hand label dispensing, hand application, bench-top, semi-automatic, fully automatic, rotary and print and apply. Linked to conveyors and product handling systems, applicators are available for the labeling of everything from tiny ampoules to large oil or chemical drums, from whisky miniatures to shrink-wrapped pallets, and from lipstick tubes to cooked meats and ready meals.

Where high-speed labeling is required, sophisticated rotary labeling systems have anything from 8 to 24 rotary table assemblies (see Figure 1.8) with cam or servo-driven bottle plate control, enabling the container to 'dance' on the platform and achieve outstanding accuracy.

These systems incorporate a rotary turret with electric motor-driven height adjustment and pneumatically adjustable top hold down pressure. Application speeds of more than 1,000 products a minute can be achieved on cylindrical, rectangular and oval-shaped bottles, using servo drives to position containers. Such machines offer multi-axis adjustment with position indicators, powered label rewind and broken label web detection, as well as label detectors that can compensate for missing

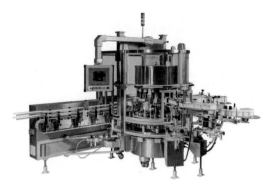

Figure 1.8 - An Accraply rotary labeling system that can be equipped with up to 24 stations for applying labels to round, square or rectangular containers

Figure 1.9 - Illustration shows a linerless label applicator (left-hand side of the picture) designed and built by ILTI

labels so that no containers are missed.

Special types of application systems have also been developed for the labeling of publications, either onto already shrink-wrapped magazines, onto bundles of newspapers or for the direct application of Front Page Notes. Such systems are regularly used for the labeling of booklets, pamphlets, brochures, magazines, trade journals, newspaper, flyers and direct mail.

LINERLESS LABEL APPLICATORS

Self-wound linerless pressure-sensitive labels are most commonly found as a direct thermally coated and top coated label material with the print surface coated with a release coating. The reverse side is coated with a pressure-sensitive adhesive. When the roll is wound up, the face stock functions as the release surface. The labels are then thermally printed, butt-cut and applied on a print and apply label applicator. They are used as weigh-price labels and traded-unit logistics labels, or with small, portable, thermal label dispensers.

A number of companies now offer linerless solutions suitable for decoration of personal care or household products. Micro-perforations allow the applied label to be separated accurately and at high speed, a system originated by Catchpoint. One of these applicator systems is shown in Figure 1.9. All other linerless systems cut and/or separate the label before application. Monoweb labels were cut and then

removed out of the web by direct application to the container.

Some systems can provide a limited number of shapes cut to the top and bottom, although the laminate material is more commonly printed, without die-cutting, at full press speed and subsequently converted into a single ply linerless label on a special machine or converting operation.

In summary, if an end user needs to apply pressure-sensitive labels to almost any kind of product shape or size, in any kind of environment, whether over-printed or not, and with or without a backing release liner, there is an application system able to do the job. The technology available today is summarized in Figure 1.10.

Very fragile or delicate products can be labeled using jets of air, without the physical contact of a roller or pad. Applicator systems employing pneumatics are used in flame-proof situations for labeling spirit or alcohol-filled containers.

Custom machines have been built to label beer kegs, car windows and make-up compacts, and for applying labels to internal membranes on air bag deployment tubes, bundling rods, and for wrapping tapered products.

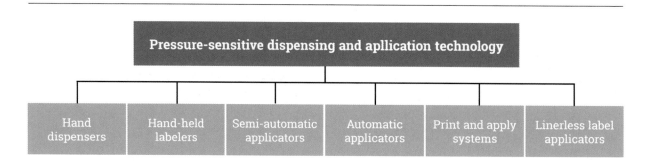

Figure 1.10 - A guide to the main types of label dispensing and application technology

APPLICATION FLEXIBILITY

The direct transfer of labels from a reel permits exact placement in front, back or neck positions, on top or bottom, into recesses and around corners in any combination, depending on the number of application heads on the machine (see Figure 1.11). Changeover of labels from one design or bottle type to another takes about 15 minutes. Any combination of labels on a three-head applicator can be changed at any time, thus providing extreme flexibility.

The latest modular labeling systems allow wet-glue, pressure-sensitive, linerless, wrap-around or sleeve labeling options on one machine.

Figure 1.11 - Different types of bottle decoration solutions now provided by pressure-sensitive labeling

Chapter 2

Understanding the basics of label application

Pressure-sensitive labels are a laminate construction made up of a label face material of paper, film or foil, a pressure-sensitive adhesive and a silicone coated backing paper or film (release liner). The release liner protects the pressure-sensitive adhesive during handling, printing, converting, die-cutting and rewinding of the labels and right up to the point of application. At this stage, the backing is peeled away from the adhesive immediately prior to its application. This pressure-sensitive label construction can be seen in Figure 2.1.

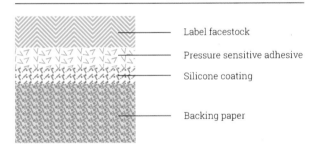

Figure 2.1 - Pressure-sensitive label construction

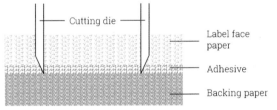

Figure 2.2 - Die-cutting through the label face material and adhesive, but not the backing paper.

Accurate die-cutting is an essential element in successful label application. The cutting die – flat, rotary or wraparound – should only cut through the face material and adhesive of the pressure-sensitive laminate (see Figure 2.2.), but not cut into or through the backing paper. The matrix is then removed to leave the desired label shapes on the web (Figure 2.3.).

The printed, die-cut and waste-stripped labels that come off the press are usually in large master reels, often with a number of label repeats across the web.

These master reels are slit and rewound into smaller, single-width, applicator-sized, reels before being despatched to the customer. This process is called rewinding and is the operation of unwinding printed die-cut labels from the press roll through a machine that slits them to the required widths, and maybe inspects them to enable faulty or missing labels to be replaced. The number of labels on each roll is counted and the labels rewound into smaller rolls of the correct size ready for end-use application, with

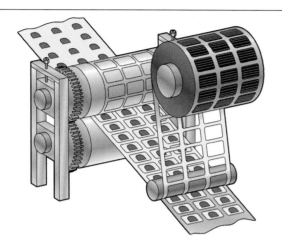

Figure 2.3 - Matrix waste being removed and re-wound after die-cutting

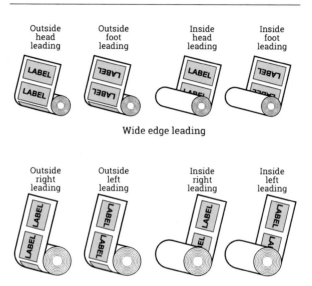

Figure 2.4 - Rewind options available with pressure-sensitive applicators

the correct leading edge for the specific application line.

Poor or faulty die-cutting may lead to problems at both the waste stripping and application stages. Too loose or too tight rewinding of the labels and the wrong leading-edge direction can cause application line problems. If there is no gap between each label on the liner, the leading edge of the label cannot be released and peel away on the beak or peel plate.

IMPORTANCE OF REWIND AND UNWIND DIRECTION

Most label application machines will only work with one label unwind direction, which needs to be known before slitting and rewinding of press rolls takes place. Pressure-sensitive labels wound in the wrong feed direction or orientation will almost certainly cause production delays and extended down time. Figure 2.4 indicates the range of rewind options used with pressure-sensitive applicators.

The rewind possibilities are as follows:
1. Outside head leading
2. Outside foot leading
3. Outside right leading
4. Outside left leading
5. inside head leading
6. Inside foot leading
7. Inside right leading
8. Inside left leading

Depending on the end-use application requirement, labels can be rewound on the inside or outside of a roll, and with either the narrow edge leading or the wide edge leading. Fig 2.5 shows labels wound on the outside, with the right edge leading. The diagram also shows how the terms label gap, label length, label width and label roll core are defined.

If both front and back labels are being applied then the rewinding specifications and requirements may be different for each.

DETERMINE CORRECT APPLICATOR REEL SIZE

As dispensing and application machines come in many different shapes and sizes it is also important to know the minimum and maximum outside diameter of the label reels that can be held in the unwind holder (magazine). The reel diameter measurement is also

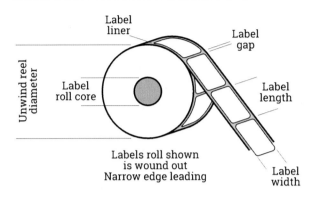

Figure 2.5 - Label rewind terminology

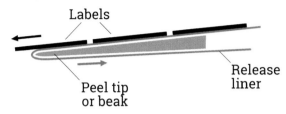

Figure 2.6 - The backing release liner needs to be pulled away from the label

shown in Figure 2.5.

The larger the outside diameter of the unwind holder, the larger the reel size that can be accommodated. Larger reel sizes on the application equipment will mean less changeover time and more production efficiency. However, the reel size must not be so large that it cannot be handled safely by the operator. Knowing the maximum outside diameter of the reels will allow the number of labels per roll and the size of roll to be calculated.

ESSENTIAL COMPONENTS OF LABEL APPLICATION

There are many different types of machines and devices for dispensing or applying labels, but they all need to incorporate a number of key features:

1. The Labeling Head, which incorporates an unwind reel holder or magazine that will carry the reel of labels, and a means of controlling the web path at the required speed through the applicator. The liner pulls the web through the labeling head and there are elaborate tension controls to ensure that the pull is always greater than the push.

2. The Applicator, with a means of peeling away the backing material (release liner) by pulling it back (see Figure 2.6.), normally over an acute metal beak, so leaving the label free to be applied to the product, either by hand or automatically, together with a means of feeding the semi-detached label forward and

pressing/applying the label to the surface to be labeled.

Extensive testing has shown that with a 2mm radius on the peel tip or beak, the optimum angle to pull the liner is 7 degrees.

The labeling head and the applicator are regarded as one single unit by most equipment manufacturers

3. The product handling system, which is designed to consistently place the product or item being labeled in the right position and speed for the applicator to do its job. Handling systems will be discussed in a later chapter.

The labeling process involves unwinding a reel of die-cut labels and then pulling the web under tension around the beak (Figure 2.7). As the backing material is taken around the sharp angle of the plate the front edge of the label peels away from the backing and the label is pressed to the surface, while the backing liner runs around the beak to be removed and rewound on the rewind take-up spindle. The aim

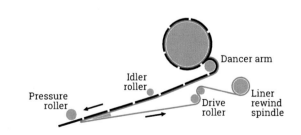

Figure 2.7 - Key elements of a label dispenser or applicator

throughout is to apply differential pressure and/or degree of wrap to achieve a consistent tension at the beak.

Many applicator systems include a separate rewind motor to maintain accurate and consistent pull tension at the beak.

UNWINDING AND CONTROLLING THE LABEL WEB

The labeling head incorporates a control module which manages the applicator functions and tells it when to stop and start.

The roll of labels to be applied is placed in the unwind reel holder or magazine, and threaded through the applicator. In many cases the label web will go around a dancer roller that helps control the unwind rate and enables a free flow of labels from the unwind and through the applicator system.

A light source and a phototransistor read the gap between labels to tell the labeler when to stop. The web path may also include an idler roller to help with a smooth passage of the web.

An adjustable tension device or shoe (Figure 2.8) is commonly located above the stripper plate or beak. This is adjusted to apply slight pressure to the label being dispensed to keep it laying flat until it reaches the plate or beak tip, ensuring an ideal pressure through a gap sensor and aiding the peeling of the label from the backing release liner.

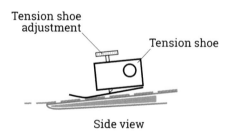

Figure 2.8 - An adjustable tension shoe

To pull the label web through the applicator system a drive shaft roller and a pinch roller are usually located towards the end of the web path. These rollers also feed the backing liner to the liner rewind spindle.

A product sensing device is used to read the presence of products to be labeled.

REMOVING THE LABEL FROM THE BACKING

One of the commonest methods of dispensing or removing die-cut labels from the web at the point of application is by using a metal beak. Depending on the equipment manufacturer this may alternatively be called a stripper plate, a tongue, a dispensing blade, a peeler or peel plate, a bar, or a peeling lip. In all cases the label web is bent around the beak or peel tip at a sharp angle, with the front edge of the label peeling away from the backing liner as a result of the liner pulling action. The liner continues on to be rewound.

Figure 2.9 - The label web is peeled around a stripper plate or beak to be removed from the backing liner. Illustration courtesy of Accraply

Stripper plates or beaks are found in a variety of configurations, including fixed plates and floating plates. Both variations can be seen in Figure 2.10.

Rigid stripper plates are usually used for very uniform products, while spring-based plates may be recommended for certain side labeling applications. Floating stripper plates, either solenoid or pneumatically operated, may be used to place labels accurately into recessed areas on a pack. Powered stripper plates need to be accurately controlled by

Fixed 90°
stripper plate

Floating
stripper plate

Figure 2.10 - Examples of fixed and floating stripper plates

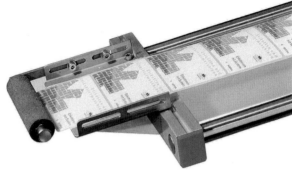

Figure 2.11 - Use of a roller to press the label to the product or pack. Courtesy of Herma

electronic timers, so that the label is pre-dispensed or maintained in contact with the pack or product for the required period.

Most systems will actually float the label onto the container surface at shallow angles. The leading edge of the label is often 'pre-dispensed' by a small length to help break the bond of the adhesive to the release coating. The 90 degree stripper plate design has a minor impact on release, as the liner is traveling marginally faster than the actual label around the guide pulley.

APPLYING THE DISPENSED LABEL

Once the labels start to be detached from the backing there are various ways of feeding them forward and pressing (applying) them to the pack, container or product. The effectiveness of the labeling operation is determined by the uniform application of pressure to ensure a positive contact of the adhesive with the surface of the product being labeled.

The three most commonly used methods of uniformly applying pressure to the applied label are:
1. Direct transfer
2. Tamp
3. Air-jet

Direct transfer application. Direct transfer, also referred to as roll-on or wipe-on, makes use of a rubber (see Figure 2.11) or foam roller and sometimes a drum applicator device to press and fix the label to the product or pack as it passes below the applicator. May also be supplemented by a brush or brushes.

Roller or drum application devices are one of the most common methods of affixing pressure-sensitive

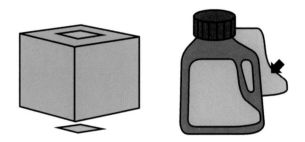

Figure 2.12/2.13 - Application of labels to the top and bottom of packs (left) and Application to the front and back of containers (right)

labels to products or containers. Ideal for flat, regular surfaces or cylindrical containers or packs, they may also be used to label concave and convex surfaces.

By employing two application heads it is possible to apply labels to the bottom and top of objects (Figure 2.12) simultaneously with only one pass through the machine. By setting the applicator heads horizontally and facing each other it then becomes possible to also apply labels to both sides of objects (Figure 2.13).

It is important that the label being dispensed is traveling in the same direction and speed as the item being labeled during the label application process.

Direct transfer label application is widely used for primary and secondary labeling and for wrap-around applications.

Tamp application. With tamp application the label is dispensed from the release backing and sucked into place and held in position on a vacuum tamp pad or plunger before being extended to the product. When the product is in the correct position, the suction is released. The label is driven to the application position by an air cylinder and applied to the product by direct pressure. See Figure 2.14.

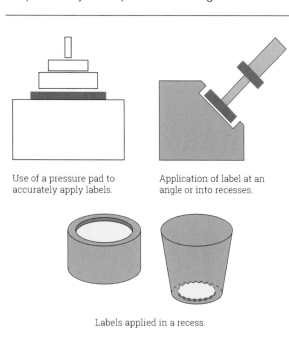

Use of a pressure pad to accurately apply labels.

Application of label at an angle or into recesses.

Labels applied in a recess.

Figure 2.14 - Examples of tamp-on label application

The result is a highly accurate labeling system. With such systems – which may also be referred to as tamp-on, pressure-pad or pneumatic plunger – it is possible to select any surface, direction or angle, up or down, simultaneously to both sides, for label application – for example to place labels at an inclination of 45 degrees.

By employing a longer plunger arm it is possible to apply labels into deep recesses or even into the bottom of something like a cup or glass. Depths up to 150mm (5.1 inches) or so may be reached. Some of these variations can also be seen in Figure 2.14.

Air jet/air blow application. This is another frequently used method of label application. The labels are stripped from the backing as usual and then, after release, are retained by a vacuum on a honeycomb grid, from where they are subsequently blown onto the product or container by applying air pressure (see Figure 2.15).

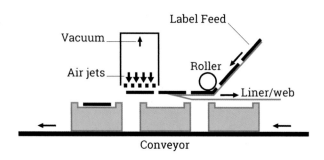

Figure 2.15 - Air blow label application

Sensors at the peel tip ensure there is no double labeling or missing labels, allowing for fast, accurate placement. Tampers or brushes may additionally be used to press the label firmly down for wrinkle-free label attachment.

Using this method it is possible to apply labels in positions that would be virtually impossible by other methods. Absolutely no damage will be done to the surface or product to be labeled.

A variation of air jet application is where the application head is built into a mechanical arm which can then be positioned to label products such as television tubes. This type of application is illustrated in Figure 2.16.

Other application options. Other methods of uniformly applying pressure to the label to ensure that it sticks with a positive contact include hybrid applicators (using two or more of the methods already described), vacuum wrap, dual-web and multi-label applicators.

It is possible to achieve great flexibility with one labeler handling a range of different products. Examples of some of the different types of dispensing and application devices are given on the next page.

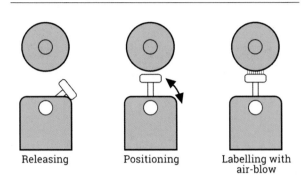

| Releasing | Positioning | Labelling with air-blow |

Figure 2.16 - Air blow application using a mechanic alarm

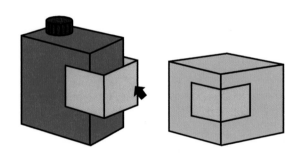

Figure 2.17 - Application of labels around corners

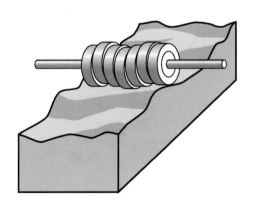

Figure 2.18 - Application onto uneven or flexible packs

With some systems it is feasible to apply labels on and around the corners of boxes, such as for medicines or confectionery box sealing, or even three panel labeling. These can be seen in Figure 2.17. A variation of conventional roller application used on some machines is the use of rollers made of knurled aluminium rings, so allowing packs with uneven surfaces to be handled, as in Figure 2.18.

Powered rotating brushes, sometimes profiled to the corner of a product, may also be used to press the label into difficult areas. Even more sophisticated methods involve wrap-around belts or reciprocating pads.

LABEL AND PRODUCT CONTROL

To ensure that labels are applied accurately and consistently to products, pressure-sensitive label applicators may be fitted with various label or product tracing/sensing/control devices such as the one shown in Fig 2.19.

Figure 2.19 - Illustration, courtesy of Herma, shows an example of a label sensing device

Label sensing or tracing devices monitor the gap between the individual die-cut labels and interrupt the web feed at the end of each label. This resets the system to receive the next start signal from the product tracing devices, which include micro-switches, photocells or spot color readers. Other controls on the web may be used to detect missing labels, the end of the reel or web breaks (see Figure 2.20 and 2.21).

With many applicator systems the actual product or

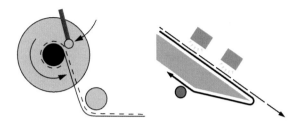

Figure 2.20 - Use of end-of-roll (left) and missing label detectors

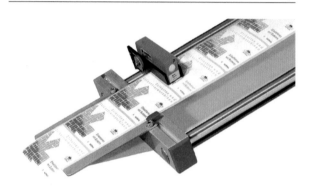

Figure 2.21 - Shows a missing label detector. Photo courtesy of Herma

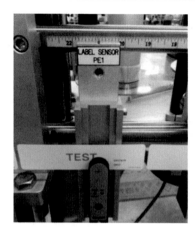

Figure 2.22 - Another type of label sensor used by Accraply

container triggers the application of the label through the use of product tracing/control devices. This means that even with erratic or irregular product flow the system will automatically compensate and only dispense a label as required.

Container or conveyor speed encoders may also feed back to the applicator so that the next label is applied at the correct speed. A fully loaded in-line slat conveyor will run faster if containers arrive with big gaps.

Many different scanning/sensing technologies are available on pressure-sensitive label applicators, from photocells with reflector mirrors, transmitter/receivers, proximity switches, spot or color readers or, for certain applications, micro-switches. For semi-automatic applications foot switches may be used.

It is also common for label applicators to be part of a larger packaging or assembly line in which they will receive the pack or product to be labeled from a previous automation stage, apply the label, and then feed the labeled item onto the next operation or stage in the process. Sophisticated handling, product orienting, conveyor belt and other control systems or devices may then be required, as well as a means of product detection and ejection (see Figure 2.23.).

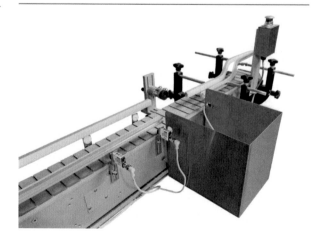

Figure 2.23 - Sensing devices used as part of a product ejection process. Photo courtesy of Accraply

Where containers are labeled empty - before filling - there are usually feedback loops with queue sensors which will slow the labeling machine if the filler stops or the bottle feed is interrupted. Accuracy is better where systems slow rather than completely stop.

Now that the basic components of label dispensing and application machinery have been reviewed, we turn in the next few chapters to how these key elements come together in different types of machines and systems.

Chapter 3

———

Manual and electric label dispensers and hand-held labelers

———

Not all businesses require sophisticated automated label applicators; many will only need simpler bench-top or stand-alone dispensing devices or basic application machines, often only hand or manually operated with little or no automation. Or may be electrically operated, and only used for light-duty applications, variable size packs, small volumes, stock office or safety labels.

———

Nevertheless, they still require some kind of machine or device that will assist in the process of easily removing each label from the backing release liner and applying by hand or semi-automatically.

For the very basic requirements for dispensing small labels or stickers in offices and on bench tops there are a number of companies that provide cardboard box dispensers already containing small rolls of blank or pre-printed labels, with the box top or side itself acting as the stripper plate. Moving beyond such simple devices, there is a wide range of economical manual or electric dispensing and hand-held labeling devices on the market. Some are portable, some bench- or wall-mounted, some are purely hand-operated, others electrically powered, but with the labels applied by hand. Hand dispensers may carry only one size or width of reel, other dispensers may hold up to five or more small reels at the same time.

These types of dispensers are the least expensive labeling machines and are usually used in relatively low-volume situations or when labels are only

occasionally required. By automatically dispensing the next label to be used, they increase efficiency over the use of manual labeling from a roll of labels.

Like the larger and more automated devices, the simpler dispensing machines still need to perform the operation of pulling the liner around a stripper plate so that the label can be easily removed from its backing and then applied by hand. Depending on the application and requirement, some dispensers are able to dispense 2, 3, 4 or more labels at the same time.

There are also easy-to-operate dispenser/applicators that can be operated by the simple turn of a handle located on the side of a bench-top mechanical labeler. This is often enough for companies looking to apply a pressure-sensitive label easily and consistently to almost any round or cylindrical product at anything up to 25-30 containers a minute.

Table-top dispensers and applicators are starting to become more sophisticated. They may be semi-automatic, electrically operated, or may incorporate a mechanical sensor, photo-sensor, perhaps a micro-switch to detect the leading edge, or an optical reader/sensor for use with transparent labels. Ideal for small labels and small work areas, table-top applicators automatically peel labels from the liner and advance and apply the label.

Hand-held label applicators of course need to be fairly light in weight so that they can be easily carried and operated without any kind of strain. They may be trigger-operated and incorporate a sensor which automatically adjusts the stroke of the trigger to the length of the label, or battery-powered. Both types are used to undertake repetitive labeling tasks in a fraction of the time required for hand labeling. In general, hand-held labelers require no special operator training or tools.

There are a number of other more specialized types of dispensing and simple application machines. One example is a wipe-on label applicator designed as a small table-top unit mounted on an in-line adjustable-speed conveyor. To use, an operator simply places the item to be labeled at one end of the conveyor where it then travels under the label head and a label is applied. The labeled item then exits at the other end of the conveyor.

Manual feed, manual place or manual push machines can also be obtained for applications such as the labeling of pouches, barrier bags, anti-static bags, vacuum bags and heavyweight poly bags. With a manual-fed bag machine for example, the operator takes a bag and places it on a plate located on the machine, pushes the bag into feed belts so as to pass through a label head which places a label in the desired location. The labeled bag then exits at the other end of the machine.

Similar types of manually-fed table-top application can be used for applying a label or seal to small boxes or cartons. Used in the pharmaceutical, medical packaging, food and other sectors requiring items packed in small boxes, the box sealing process again requires no operator training or special tools.

Having outlined the key types of manual and electric label dispensers and hand-held labelers, we now look at each type in more detail.

MANUAL AND ELECTRIC LABEL DISPENSERS

The aim of all dispensing devices is to dispense and feed an individual label forwards so that each label can be taken by hand and manually attached to a product, container, pack or surface. They are widely available in different label widths and label reel sizes and may be specific to a particular type of label or application.

Manual label dispensers. Manual label dispensers are simply operated by hand and are designed for fairly light usage. There is no automation of the operation. With manual label dispensers – including the cardboard dispensing devices mentioned earlier - all the operator has to do is pull the liner around a stripper plate or bar, so partially releasing the label from the backing liner and making it easy for the operator to pick the label off of the liner and place it where required.

Commonly used in packaging, fast food, bulk mailing, manufacturing and beverage applications, manual label dispensers may be portable, bench-top or wall mounted and dispense single labels or, with some manual devices, up to four or five different small labels side by side.

Figure 3.1 - This electric label dispenser by Start International is ideal for small labels.

Electric label dispensers. Like manual dispensers, electrically operated, semi-automated, label dispensers are designed to dispense individual (Figure 3.1) or multiple-row (Figure 3.2) labels forward and enable them to be removed from the release liner backing. However, instead of pulling on the liner by

Figure 3.2 - More than one label can be dispensed with this Start International dispenser

hand to move each label forward, label advancement now takes place when a trigger device or sensor on the dispenser detects that a prior label has been removed. The sensor then immediately closes the electric circuit so that an electric motor can dispense the next label on the roll. This action is repeated each time a label is removed.

Sensors are used to detect whether a label has been removed or is missing on the roll, and are of three different types:

1. Mechanical sensors. When a label has been moved forward electrically for dispensing it will touch a mechanical sensor that will stop the next label from moving forward.

2. Photo sensors. Photo sensors can be used for the dispensing of all types of die-cut and butt-cut labels and require no tools when changing label size or shape. There is no trigger device to adjust. Photo sensors are good for opaque labels as they can

sense the difference in color between each label and the backing liner.

3. Optical sensors. Optical sensors are the most sensitive and accurate type of sensor and are able to detect transparent, miniature, metallic and opaque labels. They are used in the dispensing and advancing of all types of die-cut and butt-cut labels independent of size, shape or transparency. No adjustments are required for changeovers.

MANUAL AND SEMI-AUTOMATIC TABLE-TOP APPLICATORS

With manual and semi-automatic table-top application machines, the product to be labeled will be manually placed by an operator into an adjustable cradle, or onto a plate or perhaps a short conveyor belt or line of rollers. The machine is operated by pushing forward the item to be labeled or by turning a handle.

Each type of product to be labeled may require a different means of manually or semi-automatically placing the product in the correct position for the label to be dispensed and applied. Some of the key types of manual and semi-automatic table-top applicators are described below.

Figure 3.3 - Bench-top hand-operated bottle labeling from Start International

1. Bottle and cylindrical product applicators. With the simplest types of bottle and cylinder applicators, the product is manually placed in a cradle and a handle on the side of the applicator turned to apply a pressure-sensitive label evenly and consistently. Such devices can be used to label almost any round or cylindrical product. Whether label application is undertaken using a manual or semi-automatic applicator, the bottle rotates during the labeling cycle.

Usually supplied as small bench-top designed devices, such machines are easy to load and use and are easily adjustable for consistent product placement. With a good operator it is possible to label up to 25 to 30 bottles or cylindrical products per minute. An example of a mechanical bottle label applicator can be seen in Figure 3.3.

It should be noted that with hand bottle labelers there may be a significant matrix loss simply positioning the labels correctly on the bottle. Depending on the level of automation, it may be feasible to save costs if sizes can be optimized through the print and die-cutting process, particularly to exploit standard web slit widths.

Next come semi-automatic label applicators for bottles and cylindrical products, but again with the products placed by hand into a cradle. Depending on the model, the operator will then either press a button, pull a handle down or step on a foot pedal to start the label operation cycle.

Once the product has been placed in the cradle and the cycle started, a spring-loaded overhead pressure arm is used to keep light pressure on the container for consistent and secure label placement. With some applicators it is possible to apply both a front and a back label during each cycle (providing the necessary labels alternate on the web). Some models may have a powered rewinder for liner waste collection.

2. Carton and box conveyor belt applicators. Bench-top applicators with conveyors offer higher productivity than a semi-automatic label dispenser and are easier for many label users to justify than in-line fully automated labelers. Designed to operate as table-top units with an adjustable speed conveyor, an operator simply puts the carton or box at one end of a conveyor (Figure 3.4.) and as it travels under or past the label head a label is applied. The box then exits at the other end of the conveyor.

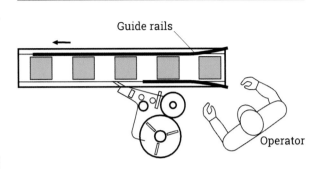

Figure 3.4 - Hand placement of cartons on a conveyor

Non-contact sensors are used to detect both the label and product. Simple push button controls with dwell timer functions ensure the label can be placed at a desired position on the product consistently and accurately. Some models have a touch-sensitive display screen where all settings can be made and viewed. It is also possible to purchase such machines with an adjustable height stand on wheels that can be rolled up to an existing conveyor line.

3. Bag and pouch applicators. Both manual and automated feed machines are available for the labeling of a range of different types of bags or pouches and many can place a label virtually anywhere on the bag. With manual fed machines the operator places each bag on a plate in the machine, then pushes the bag into feed belts where it is taken under a label head which applies the label in the desired location virtually anywhere on the bag.

Easy to use non-contact sensors are used to detect both the label and the bag. Simple push-buttons are used to control the dwell time, and motor speed if automated feed machines are used.

MANUAL AND BATTERY-OPERATED HAND-HELD LABELERS

These types of hand-held machines need to be fairly light so they can be easily carried and operated. They are used to undertake repetitive labeling in a fraction of the time of hand labeling. They enable an operator to apply labels anywhere they are needed, including assembly line labeling, batch products, palletized cartons, corrugated boxes, blister packs, PLU labels on fruit and vegetables and bar code labels.

Figure 3.5 - Manual hand-held label applicator from Start International

Manual hand-held labelers are usually trigger-operated and incorporate a sensor which automatically adjusts the stroke of the trigger to the length of the label.

Battery-powered hand-held labelers, on the other hand, incorporate a non-contact optical sensor to adjust for label length. Battery-powered machines are equipped with a touch sensitive nose, enabling the operator to simply touch the front of the labeler to the product being labeled. The label then advances and is applied in one simple motion. These labelers use either a self-contained rechargeable battery pack or direct AC power.

The more sophisticated labelers will have a rewinder for the label liner waste. In general, hand-held labelers require no special operator training or tools.

Price, 'sell by' dates, barcodes, time or date stamps, product description and tracking information can all be added using hand-held price guns (such as the one shown in Figure 3.6) and other types of hand-held labelers that can print 1, 2, or 3 lines of fixed, variable or sequential information before dispensing an application. Printing and overprinting techniques will be discussed in more detail in a subsequent chapter.

Figure 3.6 - A typical hand-held price gun labeler

Chapter 4

Automating the labeling line

The dispensing and application of pressure-sensitive labels can be undertaken on many different types of labeling machines, ranging from small hand-applied and hand-held devices discussed in the previous chapter, up to large-scale major investments in large volume, high-speed factory installations and on to completely integrated packaging, bottling, filling and label application lines.

In between these extremes are a very diverse and increasingly sophisticated range of both semi-automatic and fully-automated labeling machines that between them can be used to label almost any kind of product type or shape, in almost any position or orientation, die-cut or butt-cut, and in all kinds of label sizes and shapes.

Machines can be supplied as stand-alone units, integrated into existing production lines or installed as fully automated systems incorporating product handling, conveying and finishing units. Note however, the difference between adding applicators to existing product handling systems and the high level of sophistication and investment in fully integrated labeling machines where the machine supplier takes full control of the customer's container or package.

In all these cases, the basic principles remain the same. The pressure-sensitive label applicator will unwind and advance the label stock over the beak or stripper plate until a portion of the label dispenses and extends into the path of the oncoming product or package. As this portion engages the item to be labeled, the label advances at the same speed as the web and is pressed, wiped, tamped or blown onto the

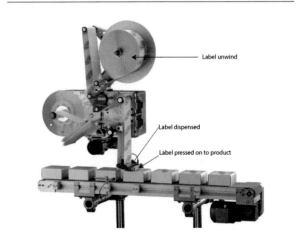

Figure 4.1 - The label reel unwinds, a label is dispensed at the beak and pressed onto the product to provide adhesion. Photo courtesy of Accraply

product to provide the necessary adhesion. These stages can be seen in Figure 4.1.

Semi - and fully automatic label applicator systems can be defined as follows:

Semi-automatic labeling machines (see Figure 4.2) speed-up the process of dispensing and manually applying product labels by partially automating the process. An operator is still required to manually place an item onto a conveyor, plate or roller, where the label will be detected and applied automatically.

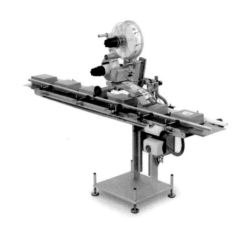

Figure 4.2 - A pedestal-mounted applicator. Illustration courtesy of Herma

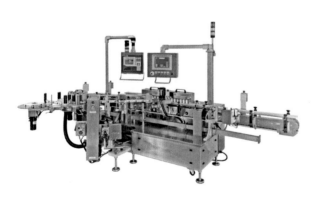

Figure 4.3 - Illustration shows an example of a fully automated Accraply labeling machine

Fully-automatic labeling machines, whether in-line or rotary systems, are used for higher volume and higher speed labeling requirements where the whole label operation needs to be automated. Products travel along a conveyor belt or other handling system and are positioned at the labeling head, where a sensor automatically activates and dispenses the label ready for application.

Fully automatic systems can be configured to suit existing manufacturing or packaging line workflows and integrated with all kinds of manufacturing, overprinting and packaging operations. These systems will be discussed in more detail in subsequent chapters.

LABEL APPLICATOR AUTOMATION

Highly automated systems use a central control module and a range of sensors to control the label dispensing and application process. Sensors are used to detect a product's orientation and location, as well as the leading edge of the label. The starting and stopping of the labeling head will be determined by a light source and sensor that read the gap between labels.

Product and labeling speeds need to be synchronized; labels and products need to be traced and controlled; missing labels need to be detected; labels or labeled products counted; and in some cases, products automatically orientated.

In addition to these requirements, both semi- and fully-automated label applicators will require a stepping motor, microprocessor controls and usually, some kind of touch-screen control. If there is a printer or coder, the control module will tell it when to operate.

The configuration of an automated applicator will depend upon a range of factors, including type of application, throughput speed, product size, shape or complexity, quality control requirements, product handling needs and the degree of integration with other packaging line operations. Elements of an automated system will include:

- Touch-screen control
- Microprocessor control/microcontrollers
- Stepping motor
- Servo-driven labeling heads
- Auto changeover labeling heads for continuous running

- Speed synchronization
- Dancer arm
- Product sensing/tracing (often using code readers with a variety of pre-printed codes)
- Spot color reader
- Web break detector
- Missing label detector
- Label scanning or sensing
- Adjustable dispensing speed
- Label dispensing counter
- Micro-switches
- Applied label counter
- Verification system if required
- Vison inspection system
- End-of-reel detection
- Auto product orientation
- Operator alarms and safety devices

Touch screen control. Touch screen technology is now widely used to control operations previously managed through mechanical dials, buttons and levers. They are also an ideal replacement for the keyboard in industrial environments and are intuitive to use, and with the right choice of touch technology, can be robust, waterproof, and hygienic.

Touch screen installations today are entirely solid state; they use touch panels on a display screen

driven by a GUI (Graphical User Interface). By changing the screen image, the area above it on the touch panel will be visually linked to a different function. This means that, as well as being reconfigurable, the system can coach a user through operations, perhaps on start-up or shut-down, or trouble-shoot when something unusual happens.

Touch screen interfaces play an increasingly important role in industrial labeling automation, because of their inherently robust nature and ability to cope with the harsh world of manufacturing.

Microcontrollers. Touch screen and touch sensing devices are supported by microcontrollers or microprocessors, together with appropriate touch screen software. Touch screen controller solutions make interfacing the touch panel to the system electronics a simple integration issue.

Micro switches. Micro switches are a key element in today's electronic automation and are used in a variety of industrial automation applications.

Stepper motor. A stepper motor is an electromagnetic device that divides a full rotation into a number of equal steps – converting digital pulses into mechanical shaft rotation. The position of the motor can be commanded to move and hold position at one of these steps. No feedback sensor is required as long as the motor or servo motors are carefully sized to the application.

The stepping motor automates the label dispensing operation and controls the speed at which labels are dispensed for application,

Label and product sensing and control (Figure 4.5). To ensure that labels are applied accurately and consistently, the semi-automatic or fully automatic pressure-sensitive label applicator is fitted with various label sensing or product sensing/tracing control devices.

Sensors range from photocells with reflector mirrors, to transmitter/receivers, proximity switches, spot or color readers or, for certain applications, micro-switches.

Label sensors are usually photoelectric sensors because they are relatively inexpensive. They cannot be used with clear labels, however, in which case capacitive and ultrasonic sensor technologies are used.

Label sensing/tracing devices monitor the gap

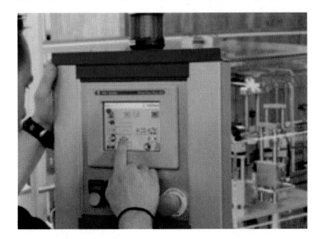

Figure 4.4 - Shows an applicator touch screen panel courtesy of Herma

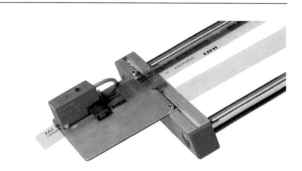

Figure 4.5 - Illustration of a label applicator label sensing device courtesy of Herma

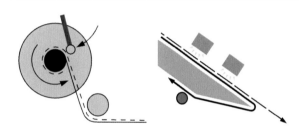

Figure 4.6 - End or unwind reel detector (left) and missing label detection (right)

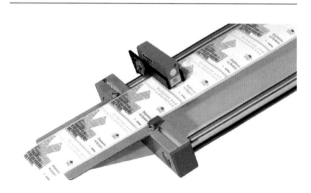

Figure 4.7 - Herma optoelectrical sensor unit

web may be used to detect end of the reel (Figure 4.6 left diagram), detect missing labels (Figure 4.6 right), or web breaks.

With many applicator systems the product or container trigger the application of the label through the use of product sensing/tracing/control devices. Product position sensors can be optical, optoelectrical (Figure 4.7) or ultrasonic. UV sensors may be used for 'invisible' sensing. Using these various types of sensors, the application system will automatically compensate for erratic or irregular product flow and only dispense a label as required.

For semi-automatic applications foot switches may be used.

Dancer arm. As the label is threaded through the applicator from the unwind reel it will typically pass around a dancer arm, a self-braking system that is able to slow the unwind reel as demand for the label decreases and so prevent the unwind reel from spinning and unwinding all the labels from the core. When label demand increases the dancer arm again enables the free flow of labels.

Figure 4.8 - An example of a product sensor used on Accraply machines

Very high-speed labeling lines with heavy reels may incorporate powered unwinds to improve tension control of the web feed.

Limit switch/photodetector. A limit switch or photodetector is used to trigger label advancement. It detects the absence or presence of a label to facilitate dispensing. Limit switches are the original

between the individual die-cut labels and interrupt the web feed at the end of each label. This resets the system to receive the next start signal from the product tracing devices, which include micro-switches, photocells or spot color readers. Other controls on the

method of label detection and can be used for most types of label application.

Photo-detectors use a beam of light broken by the label as it passes between or over the sensors.

Figure 4.9 - Operator alarm. Photo courtesy of Accraply

Operator alarm system. Operator alarms are common in high-volume and high-speed applicators. There may be visual/optical (flashing lights as shown in Figure 4.9), audio/acoustic (bell, hooter), or a combination of both. They are used as an alert or warning of a web break, missing label, end of the unwind reel, or - with a verifier or scanner - to notify poor bar code readability. Safety guards or optical screens may also be used.

An alarm alerting the operator to the end of the unwind reel can be included. In some cases the alarm system may also trigger a shutdown of the machine or the product handling line.

CHOOSING A LABEL APPLICATOR
One of the big challenges for today's automatic labeling machine manufacturers and users is the increasing variety of packages or products that are expected to run on a single line. This is particularly true in many contract packaging applications which may rapidly change from day to day or even run to run. Many contract packaging facilities may never know quite what items are to be labeled next. They therefore require their application machinery to be highly flexible.

Whether the requirement is for single product labeling or for maximum flexibility, the applicator may need to be custom built with specific product handling, environment, speed or volume requirements in mind. Either way it is likely to be a significant investment.

It is important for the manufacturer to know the environmental conditions around the application machine, such as dust, heat, cold, damp or humidity. Will electrical systems need to be carefully protected? Will the equipment need to be powder-coated?

Figure 5.0 - Operator alarm. Photo courtesy of Accraply

Then what are the volume requirements? Light duty, manual handling application? One, two or more sides to be labeled? A large volume, fully-automated application? High-speed requirements? Integrated into conveyor or product handling lines?

Questions need to be asked about service call-out times, telephone support, and on-line breakdown evaluation. These factors are all vital in assessing what equipment to install.

Ideally, end users should provide samples of the potential products to be labeled before making a final machine purchasing decision.

The next Chapter looks in more detail at how label application equipment can be integrated into conveyor or package handling lines.

Chapter 5

Integrated and 'intelligent' label application lines

Where label applicators are integrated into a completely automated and 'intelligent' packaging, handling and shipping line, labels will be applied to bottles that have already been filled and capped, or to packed cartons or integrated with a variety of different types of wrapping and production equipment, including flow wrapping, form fill and seal, bagging and shrink wrapping.

Labels may need to be applied in a variety of different positions: top, bottom, top and bottom, one or both sides, neck, angular, into a recess, partly or fully wrapped-around cylindrical containers, onto oval or unique-shaped products, corner wrapped, U-gate three panel applied, c-wrapped, top and side tamper-evident, and multi-panel labeled.

Production challenges will likely arise in the many application stages outlined above, and the key is how products are handled. Inconsistent product handling will make it very difficult to place labels consistently and accurately.

Items must be properly presented to the applicator system for labeling to take place, and for products to be taken away for onward handling. This can be achieved in one of three main ways:

1. The applicator is mounted over, or onto, an existing production line
2. The applicator is supplied as an integral unit with its own product handling or conveying system
3. The applicator is supplied on a pedestal stand or base which can be easily moved from one location to another for either semi- or fully-automatic application.

With the simpler semi-automatic label applicators, the product is manually loaded onto a table or plate at the applicator beak using either a hand-loaded jig fixture, by pushing the product along a short conveyor, or by hand feeding from another operation or a packing table. Product handling is largely manual and does not require sophisticated automation technology.

More sophisticated systems will be required as the line specification moves towards faster speeds, more automation, longer runs, application of two or more labels, the need to place labels in a recess or around corners and to control or orientate the packs. The product handling devices fitted to the labeler then need to become more sophisticated, incorporating conveyors, guide rails, belt assemblies, feed screws, hoppers, elliptical aligners, metering wheels and rotary turntables so the applicator can handle very small or difficult shapes, elliptical and rectangular containers, ampoules or vials, flat or flexible packs.

Verifiers, bar code readers, counters, visual camera/computer recognition, operator alarms, microprocessor-controlled electronics, and ever-more sophisticated operator and system controls may also be incorporated.

The three main methods for bringing the container or product to the applicator are:

Application on the production line. Many

applicators can be supplied so that the application head is mounted in position over the customer's existing production and packaging lines using a wide variety of moveable, static or adjustable brackets and angular devices to control the product and position it for application. Systems are quite flexible and can be built to be mounted above, to one or both sides and even under the production line.

Integral applicator systems. Applicators are frequently supplied as stand-alone systems with integral conveyor, container and product handling devices and with all the necessary interfaces. Both in-line and rotary systems are available.

Pedestal mounted and portable applicators. Flexibility in label application and the rapid deployment from one location to another can be achieved by the use of self-contained, pedestal, bench-operated or portable applicator systems. These can be quickly and easily moved to another production line without the need to be attached to a conveyor. They may be used in semi-automatic or fully automatic modes.

Figure 5.1. shows a label applicator positioned over an existing conveyor production line to apply labels to the top surface of a product. Portable pedestal applicators can be installed anywhere along a conveyor line as required, depending on the specific application, with adjustable leveling pads fitted as standard.

This chapter now reviews some of the most common product handling devices and controls used with automated label applicators in a wide range of end-use applications.

Conveyor systems. There are various product handling conveyor systems used to present and control the product before, during and after the label application process. Conveyors are available in different lengths, in single or multi-belt configurations, straight or curved (see Figure 5.2) and in a variety of different fabric types. The most common types of conveyors are:

- Belt conveyors
- Roller conveyors
- Slat conveyors
- Flattop chain conveyors

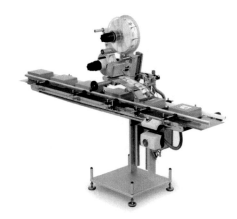

Figure 5.1 - Top labeling of products traveling along a conveyor line using a pedestal mounted applicator. Illustration courtesy of Herma

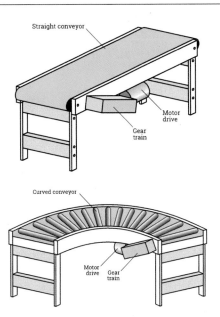

Figure 5.2 - Two examples of typical belt and roller conveyors

Of the many types of conveyors, slat (most commonly used with in-line labeling machines. See Figure 5.3), flattop chain and belt conveyors are the ones used for transporting products during pressure-sensitive label application.

There are also various specialized conveyors available, as well as side alignment or wrap-around belts, top hold down conveyor belts, stabilization belts and rotary spacers.

Conveyor components. Conveyors used for product handling in label application lines will usually be on legs, either fixed or adjustable (casters

optional), incorporate chains or belting, side supports and guide rails (see Figure 5.4) and, depending on the application, may also contain star wheels and scrolls.

With a simple conveyor, guide rails will control the product reasonably well if the product is round, square, or rectangular. Oval shapes, such as shampoo bottles, cannot be controlled correctly without additional components.

Guide rail options include single or dual rails with adjustable in-and-out as well as up-and-down mounts (see Figure 5.5). They are used to direct the product through the system and ensure alignment at the peel tip for proper placement of the label. The type of guiderail to be used must be selected after testing with the labels to ensure that there is no scratching or

Figure 5.3 - A slat conveyor. Photo courtesy of Universal Products

Figure 5.5 - Guide rails can be adjusted in and out as well as up and down. Photo courtesy of Accraply

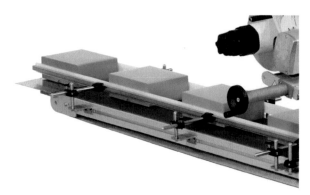

Figure 5.4 - Use of guide rails on a conveyor line to control the product. Illustration courtesy of Herma

marking.

A sensor detects the product as it comes down the conveyor and triggers the application of the label at the proper time. Product handling must be tightly controlled to ensure accurate label placement.

Feed screws, also referred to as an In-feed Screw Assembly, enable products to be accurately metered, spaced and/or orientated on the conveyor line in preparation for label application. Products will be consistently in the same position and at the same rate or speed as the conveyor at the end of the screw. Single screw models are used for round products (as shown in Figure 5.6 top diagram), while dual screw models (Figure 5.6 bottom diagram) are available to accurately align and meter oval, elliptical, irregular and square products. Product can also be captured by a top hold down belt while they are still in the screws.

In-feed screws are mainly used in primary labeling applications, such as consumer products, pharmaceuticals, home products and cosmetics.

Feed screws (Example shown in Figure 5.7), usually custom designed for each product, precisely control pitch and offer maximum product throughput at any given conveyor speed. In some cases, a single screw profile can be designed to accommodate multiple product configurations. Most applications require multiple screw change parts.

Figure 5.7 - Single in-feed screw metering bottles on an Accraply line

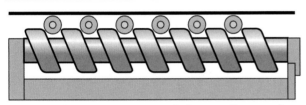

Round bottles on a conveyor being equally spaced by a single worm drive

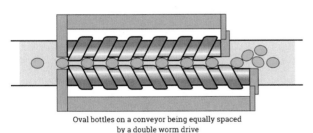

Oval bottles on a conveyor being equally spaced by a double worm drive

Figure 5.6 - Diagram to show the principle of single and double in-feed screws

Figure 5.8 - Metering of bottles traveling between guide rails on a conveyor line. Illustration courtesy of Universal Products

Metering wheels are used to space products along the conveyor line just prior to label application. (This process can be seen in Figure 5.8). Metering wheels do not provide product orientation and are mainly used on slower to medium-speed application lines.

In general, the higher the application speed, the more unstable the container, hence the need to move

on to more accurate orientation and capture using top hold-down belts or a rotary platform (see also Figures 5.9 and 5.10).

Star wheels are another type of handling device that is used for spacing products. They are usually found as a single or double-stacked wheel with a number of pockets cut to the diameter or shape of the container being labeled. The star wheel rotates in unison with the conveyor to space products for labeling.

Different sized wheels can be easily put in place and will only require adjustment of guide rails.

Spacing belt assemblies consist of two opposed timing belts that grip products from the sides and are used to space products at the in-feed end of a system. They are able to provide up to ten times more holding force than a typical metering or spacing wheel. May also be used to transfer products from one conveyor to another.

Top hold down or stabilizer belts, as shown in Figures 5.9 and 5.10, are used to stabilize products after they have been spaced or orientated for the label application process and are an integral component of a product handling assembly. They normally feature a moving belt on an adjustable column above the product conveyor used to hold products stable and maintain orientation while exposing all sides of the product for labeling. It is vital that the top hold and conveyor speeds match. The top hold must also be parallel to the conveyor. If

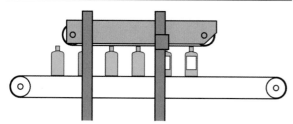

Unsteady bottles on a conveyor belt being held in place
by a top conveyor

Figure 5.9 - Shows the principle of a top stabilizer belt

Figure 5.10 - Top stabilization belt on an Accraply machine

angled the pressure can increase or decrease, affecting speed and accuracy.

As with guide rails, the top hold down moving stabilizer belt feature can be adjusted vertically to accommodate various height products.

Orientation devices, such as those illustrated in Figure 5.11, are mainly used in slow to medium speed label line applications in which it is necessary to orientate cylindrical-shaped products that may have a handle or seam so that they are positioned in the correct relative position each time prior to accurate label placement.

Elliptical aligners are used to center and orient/align an elliptical, oval or non-cylindrical product on

Figure 5.11 - Shows the use of an orientation device to correctly position products prior to applying labels. Illustration courtesy Universal Products

the conveyor top prior to application. Often uses two soft foam belts, with a durable outer skin, to prevent marking the products. The product to be labeled travels between the aligner belts to be positioned.

Wrap belt assembly. Used with some label applicator and product handling systems to rotate or spin cylindrical products (typically a bottle) simultaneously to label application, so completing and finishing label adhesion to the product. The assembly consists of a coated timing belt and an adjustable padded back-up plate. They can also be used with square and rectangular-shaped products.

Foam rollers of different diameters and widths

may be used as a secondary application device to ensure complete label adhesion to the product.

Modular options. Modular options include spacing wheels, product in-feed scrolls, in-feed and out-feed rotating tables.

Speed control. With label line integration it becomes important to track the varying speed of a product or web as production speeds ramp up and down. A bespoke speed control system, coupled with the stepper motor technology used in the label dispenser, is designed to ensure accurate labeling across the whole speed range of the production line.

Counting and tab inserting. With some integrated labeling lines it may be a requirement to

Figure 5.12 - Examples of foam rollers, courtesy of Accraply

incorporate a counting and tab inserting machine that counts exactly sheets of cartonboard, paper, plastic, inserting tabs if required.

Friction feeding. Integrated labeling lines used for placing labels on many kinds of flat products, such as greeting cards or media products, may need to incorporate a friction feeding system (Figure 5.13), where the lowest product lies on a friction belt, with a braking roller preventing the next product from being

pulled through. The opening between the belt and the braking roller can be set to the appropriate product height to give efficient product changeover.

Counting can also be incorporated, as well as missing sheet/product detection.

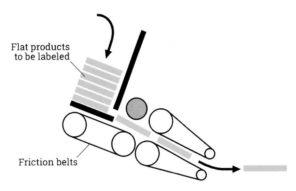

Flat products to be labeled

Friction belts

Figure 5.13 - Diagram shows the principle of friction feeding

Wrapping. Labeling machines can be incorporated into wrapping lines adding labels to, for example, medical or toiletries products, DIY parts, stationery, cosmetics and greetings cards.

Product handling flexibility. Label applicator product handling systems are highly flexible and can be arranged to label one or both sides, top or bottom, around corners, into recesses, and onto to all types of shapes and products.

Applicators may be stand-alone or rotary systems incorporating all the necessary product handling technology, or can be fitted over, under or into many different points in a customer's completely conveyorized existing product handling line. Such systems may change direction, go around corners, enable multiple lines to be fed into a main conveyor, have several lines merging onto one conveyor, and may incorporate an in-feed hopper and out-feed accumulation conveyor.

Chapter 6

Overprinting, coding and marking on label applicators

Label imprinting is pretty much a standard option for today's applicator technology and can be regarded as a key driver of pressure-sensitive labeling. A wide range of label users need to add some kind of last minute product or shipping information, and there are various ways that this can be done, depending on whether the labeler is hand-held, semi-automatic or automatic - or even on or over a conveyor or other handling system.

With hand-held labelers the requirement will be for a relatively simple, lightweight, mechanical printing unit with an inking roller or pad that can print up to three lines of limited information immediately prior to dispensing and applying the label.

The inclusion of a thermal, hot-foil, inkjet or laser printer with a label applicator head creates a labeling system that can electronically or mechanically add overprinting, coding or marking information on-demand to a pre-printed or blank label, or generate an entire unique printed label run at the point of application.

Such facilities are most often used to print barcodes, batch codes, lot numbers, time or date stamps, price and weight details, and tracking and trace information. The print and apply principle is also used widely to add unique labels to shrink wrapped traded units, chemical drums and complete loaded pallets of products, together with a whole range of turnkey applicators designed and built for specific industries or end-use applications.

HAND HELD PRINTER LABELERS
Hand held labelers with the capability of mechanically coding/printing one, two or three lines of information before applying the label are used by high street retailers, factory outlets, department stores and cash and carrys around the world for their daily price and seasonal price marking requirements, 'sell-by' or 'use-by' data, egg box labeling, simple coding, dating and batch labeling needs.

They are also found in many different industrial and manufacturing applications for SKU control, department identification, product identification, inspection, test and ISO 9001 : 2000 accreditation, coding and batch numbering operations found on production lines, in laboratory research applications or for basic job processing and sequential numbering requirements.

Another key area of application for hand-held printer labelers (See Figure 6.1) can be found in the field of distribution and shipping, warehousing, storage, packing and logistics management and control. Durable machines and labels are also available for use in outdoor applications, such as garden centers and DIY. Virtually any retail, manufacturing or packing operation can benefit from the versatility of these simple to operate machines.

Depending on the manufacturer or supplier, and on the particular end-use application, hand labelers of this kind with print capabilities may also be known as pricing guns, label guns, labeling guns, coding guns or sequential number labeling guns (sometimes also called numbelers).

Figure 6.1 - A typical hand-held printer labeler

There are many different hand labelers on the market capable of adding some level of ink coding and printing. The most widely available are:

One-line label guns. These are usually capable of printing between 6 and 10 alphanumeric characters/digits in a single line on plain or pre-printed labels. One line labeling guns are ideal for the retail trade or simple coding, dating and batch labeling needs. Also used for batch numbering.

Two-line label guns. These can print two rows of up to 10 alphanumeric characters each (20 characters/digits in total) onto labels, again either using plain or pre-printed labels.

Three-line label guns. These print up to three lines of data on a plain or pre-printed label. The amount of data that can be communicated (up to 14 characters per line) allows for extremely cost-effective and complex bulk labeling solutions. Three-line

labeling guns are ideal for production control and quality control purposes.

By combining, for example, three ten digit bands with up to 40 characters, the quantity of information that can be conveyed is enormous. Alternatively, customers can have 'data bands' installed with pre-defined specific names, such as location, date or production line codes.

Sequential number labeling guns. These are much like normal labeling guns but each time the handle is squeezed the next number in the sequence is printed. A dial enables the desired print to be achieved, from a simple date to an alphanumeric part number. They typically print from 0 to 9999, and may also have a number of fixed characters.

The information that can be printed on hand labelers is dependent upon the print head bands, which can have a range of pre-set characters. Inking of the alphanumeric or numeric characters is achieved with either inking rollers or pads. A range of ink colors are available. The ink rollers and pads are easily replaced when the print quality starts to deteriorate.

Hand-held labelers with print capabilities are easy to operate and users become proficient within minutes. A simple dial enables the desired print to be achieved from a date to an alpha numeric number. Each time the handle is squeezed the next label in the sequence is printed, dispensed and applied. They are normally very robust and require minimal maintenance.

PRINT APPLY LABEL APPLICATORS

Print and apply label applicators incorporate a printer and software to print or overprint labels on demand with details that may include a batch number, production date, contents, weight, price and transit data as well as simple graphics for product branding.

Print and dispense or print and apply labels are widely used by retailers to print weight and price information at or near the point of sale onto products as diverse as meat, fresh fruit and vegetables and other products not pre-packed prior to arriving in the supermarket. Often the labels will include a bar code which will be scanned at the supermarket check-out.

Print and apply labeling equipment is also commonly found in warehousing and distribution to provide delivery, batch and other transit information which can be used to track, trace and confirm that the product has arrived where it should, on the correct date and in the right quantity. These labels may be printed directly on to the outer transit container or onto a label which is then applied to the transit container. Information can be printed in both standard type and as a bar code.

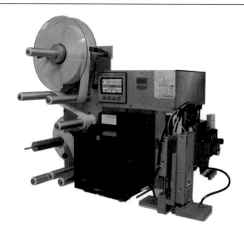

Figure 6.2 - A printer applicator that can be used with Sato or Zebra print engines combined with PLC control. Illustration courtesy of Accraply

On-demand print and apply labelers, also called 1-to-1 labelers, apply the label that has just been printed. The print engine and print head are right at the dispensing edge. These machines – whether semi-automatic or fully automatic – are used for applications where label data changes regularly, such as in distribution/logistics operations, sandwich pack labels, pharmaceutical/clinical trials applications, pallet labeling, or anywhere where products arrive in a random or non-linear order or need time and date marking for tracking purposes.

Print heads used with print and apply label applicators are typically thermal transfer with thermal ribbons suited to the label substrate, print surface or end-use application.

There are also batch print and apply labelers that feature a thermal transfer print engine located away from the dispense edge, so that the two tasks can

work independently. In this case, labels are printed into a tension controlled loop, giving opportunities for high-speed direct wipe-on application which is not limited by the speed of the printing. Typical batch applications include bakeries and other fresh food labeling, outer case labeling, and price and promotional labeling.

Tracking labels may also be applied at the beginning of a manufacturing cycle, for example when tracking a complex engineered component through a fully automated machining and inspection system.

PRINTER SOFTWARE

To fully utilize print and apply label applicators, it is necessary to either transmit data to them or to create at the printer a label format and content file so that the labels can be printed with the required information and quantity.

A number of label design software packages are available, such as BarTender, LabelView or NiceLabel, which can send label information directly to the printer. This usually involves using the software's printer driver to convert the data into a format that the printer understands.

Another common approach is to design a label file that is formatted in the programing language of the printer and send it straight to the label printer, with no label software involved. This is often used in large enterprises, where the company's ERP software is set up to send a formatted file to the label printer when a label is required.

Each printer manufacturer has its own proprietary programing language. By virtue of having the largest market share in barcode printers, Zebra's ZPL programing language has become very much a standard for developing label files. The other printer manufacturers have their own programing languages: DLP for Datamax and IPL (Fingerprint) for Intermec are just two examples.

For end users, getting locked into the language of one printer company makes it hard to switch brands: the aggravation of reformatting a whole lot of label files can be too much.

To try and overcome this, label print engine companies have developed emulation firmware allowing their product to work with label files designed for other brands of printer. This enables a user to send, for example, a Datamax DPL file to a Zebra printer. The printer software will then convert this to compatible Zebra language and produce the correct label.

PRINT ENGINES

A number of print technologies can be added to applicator lines to provide last-minute text, codes, date information, pricing, weight, volume, graphics or personalization. The print technology used depends on:

- Label size and dimensions
- Print quality
- Amount of type or data to be printed
- Size of print characters – very small/large
- Bar codes required
- Printing speed to match line speed
- Type of applicator
- The size, dimensions and label positioning on the product being labeled
- Ease of set-up and changeover
- End-usage conditions - hot, cold, outdoor, product resistance, etc.

All these factors will need to be taken into account when deciding on the most appropriate print technology. It should also be remembered that print heads and items like thermal papers, hot foils, thermal ribbons, inkjet cartridges and inks are consumable items that need to be changed or regularly purchased. Simply neglecting to clean a print head can potentially halve its life and therefore add to print

engine running costs.

The range of available OEM print engine technologies are shown in Figure 6.2.

Thermal direct uses a heat-sensitive, chemically coated label substrate into which the print images are 'burned' at standard printer temperature and pressure settings. Heating of the thermal material is with a print head consisting of many miniature heating elements distributed along its printing width, which turns the image areas dark to create the required printed label (See Figure 6.3). The elements are selectively heated by

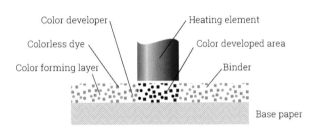

Figure 6.3 - The special heat-sensitive chemical coating on the label substrate darkens under the action of the heating elements.

pulses of energy which create points or dots of black (other colors are possible with appropriate paper) in the thermal coating, and these dots create the image held in the computer memory.

Direct thermally printed images can be variable text, bar codes, or diagrams used for frozen or fresh food labels in supermarkets, industrial bar-coded

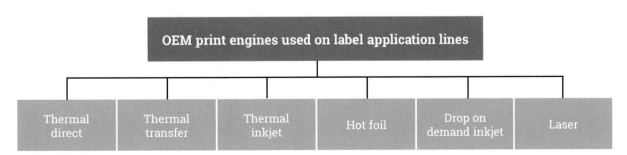

Figure 6.4 - OEM print engine technologies

labels and tags. It may also be called chemi-thermal material.

Thermally coated substrates are of course heat sensitive, so cannot be used for applications where the label may be exposed to sun, heat or abrasion. However, there are thermally coated substrates with different reaction temperatures that can be used to minimize or overcome some types of heat sensitivity issues.

Thermal transfer printing is a very similar indirect printing process to direct thermal printing, where a heat sensitive ribbon replaces the heat sensitive thermal paper. Very smooth and receptive conventional paper, or a suitable film, is printed by means of an electronically-controlled printing head which transfers variable data – including bar codes – to the label, ticket or tag face material via a heat-sensitive printer ribbon placed between the printing head and the face material.

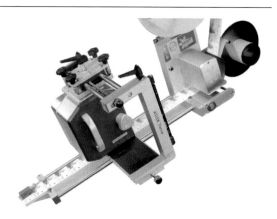

Figure 6.5 - Thermal transfer printing head prior to label application. Illustration courtesy of Herma

The thermal transfer printing head (Figure 6.4) makes use of very small resistors or elements which are arranged across the printing width, which are selectively heated and cooled. The heated elements come into contact with a thin film one-pass ribbon, which carries a heat-activated ink or coating on the underside. By rapidly heating and cooling the resistor dots on the print-head, the required character or

image is created and so transfers the selected heat-activated ink coating from the film carrier to the substrate according to the pattern or shape of the heated elements.

Thermal transfer printing is used for printing variable batch codes, date codes, sequential numbers, text, diagrams and bar codes onto pallet, carton or box end labels. Typical end uses are for warehousing and distribution, bakery labels, DIY, industrial labeling, and for a variety of tickets and tags.

Thermal inkjet printers make use of electrically heated cartridges. Each cartridge contains a series of tiny chambers, each containing a heater. When the ink is heated by a pulse of current passing through the heating element, a bubble is formed, causing a large pressure increase and propelling a tiny drop of ink onto the label substrate. Once the droplet has been ejected the bubble in the chamber collapses, the chamber refills and the whole process repeats.

The technology offers high resolution coding for the food, beverage, tobacco and cosmetics industries and may be found in intermittent or low volume production as well as higher speed coding.

Hot foil printing is a dry printing process which uses very thin aluminum foil in a variety of metallic colors – such as gold, silver, red or blue – rather than inks from which to print. Hot foil printing is achieved

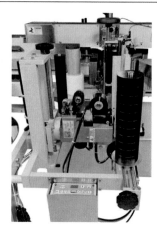

Figure 6.6 - A Norwood 50/30 hot foil coder used to print an expiration date on a label. Photo courtesy of Accraply

by transferring the colored metallic pigment coating from a ribbon of plastic material known as the 'carrier' onto the surface of the label material using a hard printing plate which bears the image to be hot-foiled.

The transfer is achieved through the application of heat, pressure, and the length of time the heated coating area is in contact with the substrate – known as the dwell time.

The balance and control of these elements is critical and must be individually calculated for the surface to be printed, and the type of ribbon or foil used. In particular, accurate pressure setting is vital. Too high a pressure may lead to line downtime with glassine liners.

The hot foil process uses relatively simple equipment and can print on a wide range of surfaces. On-line hot foil coders are used to code expiry dates, product identification information, lot and batch numbers, or may be used to provide a luxury (metallic) look on many cosmetics, toiletries, health and beauty labels, on wines and spirits labels and in other higher added-value label applications.

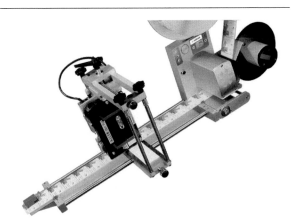

Figure 6.7 - Hot foil coding of labels on a Herma label applicator

Hot-melt drop-on-demand inkjet uses solid, wax resin-based hot-melt ink blocks that are instantly touch dry after printing. The ink blocks are solid at room temperature (and can be picked-up and handled) and are melted internally by the system as required for printing. The ink, now in liquid form, is expelled from the nozzle by piezoelectric technology. The ink solidifies and hardens immediately on contact with the label substrate.

Drop-on-demand inkjet printing ejects ink droplets from a nozzle to the substrate on an 'as needed' basis, instead of a 'continuous' basis, eliminating the need for an ink recirculation system. DOD print heads are made up of many nozzles, each capable of emitting up to 30,000 finely controlled droplets per second.

The technology provides high quality and high speed digital printing of variable data, serialization, traceability and offers fast job changeover. It is found in applications including confectionery, frozen foods, dairy, personal care and pharmaceuticals.

Laser technology is used for high-speed printing of large coding areas and can be found in the food, beverage, cosmetics, toiletries and pharmaceutical sectors. Laser printers (see Figure 6.7 for an example) print image characters using a laser and a photosensitive metal drum. The laser imprints the image on to the drum, changing the electrical charge. Toner then adheres to the charged areas on the drum and is then transferred to the label material.

Figure 6.8 - Shows a 10 watt laser printing on to Catchpoint linerless labels

Laser printers require larger cartridges as well as the printing drum and are therefore typically larger that inkjet printers – and require more space.

INVESTMENT DECISION

Before investing in a label applicator print technology, a total cost of ownership study should be undertaken, including consumable inks, ribbons and spare parts along with an assessment of the label substrate/media. Using cheap label materials can reduce the life of print heads, for example. The quality of the overprint information should also match the quality of the pre-printed label, while the accuracy, quality and readability of bar codes is essential and needs to be carefully evaluated before investment takes place.

If budgets are determined in separate locations, the total applied print cost and code readability evaluation may often be missed. Failure of bar code readability at the point of sale can certainly be very expensive.

Developments in print technologies are continually taking place and it is essential that application machine producers and end-users keep abreast of new technology developments.

WEIGH PRICE LABELING

Weigh Price Labeling systems are used in grocery store backrooms for the labeling of fresh meat, poultry, fruits, and vegetables, prepared meals, convenience foods, fish and cheese. Flexibility, speed and reliability are all critical factors in selecting a labeling system.

Packers and importers of foodstuffs must ensure that packaged goods are labeled with the quantity. Key requirements are that information printed must be visible, accurate, easy to read and understand; be labeled in the same field of vision as the name of the food; not be misleading; and not be able to be damaged.

The rules on quantity labeling of pre-packaged foods changed in Europe in December 2014. Businesses now have to comply with the requirements of the EU Regulation on the Provision of Food Information to Consumers (1169/2011). This states that all packaged foods of 5gm or 5ml or more are required to comply with FIC and that the packages must display

- The net quantity (in g, kg, ml or l) directly on the package or attached label
- Drained net weight and net weight for foods packed in a liquid medium.

As with the print and apply systems already discussed, weigh price labeling equipment comes in a range of manual, semi-automatic, fully automatic systems. Standard thermal label printing solutions are used.

Typically, weigh price labeling applications require input and data from an in-motion checkweigher to populate the labels with the correct information, weight and lot number. Usually, weigh price labeling is located toward the end of production line operations.

Apart from incorporating a printhead and a weighscale interface, weigh price labeling systems are commonly supplied with a touchscreen controller to control the weighing and labeling operations. The latest generation of open architecture equipment can also collect real time data on packaging line productivity and status of customer orders.

The more sophisticated weigh price application machines have systems able to recognize different cuts of meat, for example. Differential prices per kilo are calculated and five or six applicators place printed labels onto different height packs.

Chapter 7

———

Linerless labels and their application

———

Conventional pressure-sensitive labels consist of a face material, an adhesive and a siliconized release liner backing that functions as a carrier for the label and protects the adhesive layer during handling, printing, finishing and right up to the point that the label is ready to be applied.

———

Applicators and labelers for conventional pressure-sensitive labels need to peel away the backing release liner for each label to be dispensed and successfully applied to a container, product or pack. The liner then needs to be disposed of as waste. Conventional pressure-sensitive laminated labels like this produce the highest waste levels of any packaging component with over 50 percent lost during conversion and final end-use application.

Linerless pressure-sensitive labels enable the release liner to be eliminated. Either a release coating is applied to the label facestock itself, which prevents the adhesive on the underside from sticking to the label below, or the original clear filmic liner is transferred over the printed face stock to both protect the print and reposition the silicone release coating to the outside of the printed face. The original waste liner is now part of the applied label, so eliminating waste collection and disposal at the end user.

Eliminating or using the release liner in a more sustainable format means that savings can be made in face material costs, since a degree of stiffness is no longer required to peel from a liner. In addition, the number of labels on a roll increases dramatically - which means fewer roll changes - no matrix waste needs to be disposed of as there is no peel gap between labels, and there is a significant reduction in waste.

There are obviously distinct advantages to going linerless. Linerless labels contribute to environmental sustainability, increase production capacity, reduce applicator downtime, decrease inventory space and enable the label user to cut out the stages of liner waste collection and disposal. That can be quite substantial for them.

Linerless pressure-sensitive labels are not new of course. They first came to the fore in the early 1980s, when Waddingtons in the United Kingdom developed a technology and coating system called Monoweb to produce linerless labels which found application in companies like Heinz. They were used with a specially designed applicator system that die-cut and applied the label in one pass.

Today, linerless labels are most commonly found in the form of pressure-sensitive labels for the blank label industry, as well thermal labels used in print and apply weigh-price label dispensers and applicators. They are also popular in market sectors such as food and logistics, but have been slower to make a significant impact in the wider label markets.

That situation is now beginning to change as more countries start to tax liner waste as packaging material rather than industrial process waste, and it is becoming more complicated to dispose of.

A number of companies offer proprietary linerless technology and applicator systems for both primary and secondary product decoration. These companies include Catchpoint, Ravenwood Packaging, ETI Converting, ILTI and Ritrama CORE, which will all be discussed later.

MANUFACTURE OF LINERLESS PRESSURE-SENSITIVE LABELS

There are two conversion methods used today for linerless labels, and these are described below:

The silicone coating method. The manufacture of linerless labels starts with a converter either printing the desired images on to the face of a web of paper or film material, or by coating a direct thermal material. When printed by the converter this is undertaken in the normal way on a label press by flexography, litho or any other process, and no special printing equipment or technology is required.

A silicone face coating is then applied to the face of the pre-printed web of labels and then a pressure-sensitive adhesive is coated on the reverse of the web, with the label web re-wound on itself. The silicone face coating prevents the adhesive sticking to the facestock when it is wound up. The linerless labels resemble a large roll of printed sticky tape. This can be seen in Figure 7.1.

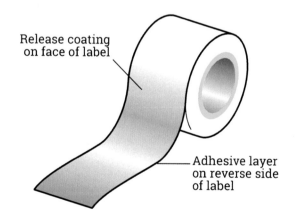

Release coating on face of label

Adhesive layer on reverse side of label

Figure 7.1 - Linerless labels can be regarded rather like a large roll of adhesive tape

With these types of linerless labels the shape of the label is limited because there is no liner to hold the label in place when it is dispensed and cut with a built-in cutter. In this case linerless label users are limited to square or rectangular shapes – frequently perceived as one of the drawbacks of the technology.

Depending on the proprietary linerless applicator technology being used, the label web may incorporate micro-perforations enabling the label to be separated at a tooling station in the applicator.

Linerless from a laminate. This new development is focussed on the use of filmic materials for product decoration, the most rapidly growing part of the pressure-sensitive label market. Again, an original laminate is printed conventionally, but now after printing, the clear film liner is transferred from the adhesive side to be over-laminated over the printed face. This positions the original release coating to now provide the coating in the linerless format. The major advantage is that this new combination of films allows much reduced face film callipers, particularly as linerless labels do not required the same stiffness characteristics – the peel plate or beak are replaced with cutting or alternative label separation systems.

Some linerless systems allow limited design shapes to the top or bottom of the printed labels. And by printing on an ultra-clear material it is possible to reproduce virtually any shape or size of label possible with conventional pressure-sensitive labels.

One of the distinct advantages of linerless labels is that the release coating is applied to the face of the label. This means that, in addition to preventing the adhesive from sticking to the face beneath it, the release coating also acts as a protective coating for UV, moisture, and chemical resistance, all adding to a label's longevity, as well as protecting the printed image and improving durability in demanding applications.

One of the common misconceptions among label converters is that special printing and converting equipment is needed to print linerless labels. This is not the case: the labels are made using conventional printing and converting processes. However, it's the point when they are applied (or re-wound prior to application in some cases) where additional or adapted equipment comes into play.

LINERLESS LABEL APPLICATION

Linerless technology ticks many sustainability boxes. It has been used with linerless print and apply label applicators for some years and is now commonly viewed as a 'greener' alternative to conventional self-adhesive print and apply label applicators, eradicating waste, reducing storage space, offering more labels on the roll, and increasing productivity.

Linerless print and apply label applicators normally include a thermal transfer printer to print data onto the linerless material, a cutter to cut the material after it has been printed and an application head of some form to apply the printed label to a product. New systems using lasers may reduce total applied costs.

Moving beyond the widely-used linerless print and apply equipment, improved coating technology, new variations of linerless production and application, and continuously developing linerless applicator equipment all look set to have a big long-term impact in key prime label sectors over the next few years, particularly in food packaging and the health and personal care markets – especially as linerless labels become less and less restrained by size and shape.

Some multinational end users and retailers are starting to push the technology forward in the US and Asia, while in Europe a number of retailers are switching the presentation of some fresh produce trays, ready meals, meat and fish packs, into linerless adhesive sleeve formats, usually with variable weight and price data, where they are beginning to replace both conventional pressure-sensitive labels and carton board sleeves.

Linerless is also poised to make further inroads into the beverage sector with the labeling of glass bottles and jars, while continuing to build on existing thermal transfer print-and-apply growth in the logistics, and data labeling sectors. Combined with the expansion of used liner recycling schemes, the latest linerless prime and decorative label technologies will undoubtedly make a big contribution towards helping self-adhesive labeling become cleaner and greener in the coming years.

We now turn to an examination of linerless technology by manufacturer.

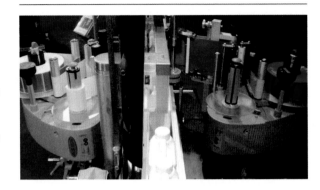

Figure 7.2 - A two-headed label applicator applying conventional labels on the left and Catchpoint linerless labels on the right

CATCHPOINT

Working closely with established material suppliers, printers and labeling machine manufacturers, Catchpoint was one of the first companies to develop a really practical linerless application system with none of the efficiency and safety risks seen with the earlier label cutting alternatives.

In operation, all Catchpoint related linerless application systems use unique micro-perforation technology. System benefits include:

- A comprehensive range of application options across the whole speed range, including ILTI proprietary rotary units, with an emphasis on the growing film label market offering thinner face materials to lower costs
- A simple applicator change kit which can be merged into existing lines to minimize investment costs
- A basic label application system that uses micro-perforations accurately imposed by conventional die-cutting systems
- A range of shapes which meets many existing label profiles and encourages printed shapes in clear films
- An easy switch for elaborately shaped labels to exploit the availability of 'LeanLiners' in the same applicator change kit. Application of 12 micron PET liners is proven
- This dual function encourages change with a

practical materials approval process and recognition that linerless label supply capacity will require time to develop
- Eliminate web break risks still associated with conventional glassine liners
- Catchpoint will shortly add an end of line 'Traded Unit' label printing and application system exploiting the efficiency and reliability of laser technology.

Investing in Catchpoint 'LinerLess or LeanLiners' is claimed to offer label users 2-3 percent line efficiency gains with a unique degree of flexibility.

ETI CONVERTING EQUIPMENT
ETI Converting Equipment, the developer of Cohesio equipment, has been working for some years to find solutions to the issue of release liner waste. Their Cohesio press can manufacture pressure-sensitive linerless labels right through from the raw material to the finished label product. The technology offers converters the ability to siliconize, print, adhesive coat, print again and die-cut at speeds up to 500 ft/minute.

More recently, ETI introduced a linerless labeler which can be retrofitted to existing applicator systems and can be used to transform the linerless 'tape' web into the same range of shapes and sizes allowed by conventional pressure-sensitive label technology.

Figure 7.3 - ETI linerless labeling solution

Being servo-driven it can adjust itself to create each label shape and size.

RAVENWOOD PACKAGING
Ravenwood Packaging is a company that manufactures and sells linerless labeling machinery to label printers as well as to retailers and packers. The vast majority of Ravenwood linerless labels produced are being used in the fresh food markets. Examples can be found in the packaging for chilled meat, fish and poultry (see Figure 7.4.), as well as plastic and glass bottles and jars.

Figure 7.5 - Nobac 500 linerless label applicator

Ravenwood's technology pairs a Comac 500 coater with a variety of the company's Nobac applicators. Both machines work together to provide a completely linerless labeling workflow.

The Comac 500 machine is used instead of a slitter rewinder, coating the labels with silicone release on the front of the labelstock and with an adhesive on the back at up to 450 feet/minute. It is specifically designed to work with the Nobac applicator.

The Nobac line of linerless label applicators is available in three different models, the 500, 400v, and 125. The Nobac 500 is an in-line machine developed for retailers and packers and is the most versatile in terms of what types of constructions it can produce. It is capable of applying sleeves in five formats: top, top and side, top and two sides, C-wrap, and full wrap. In addition to sleeving fixed weight products, it also has

Figure 7.5 - Nobac 500 linerless label applicator

Figure 7.6 - The ILTI Proper-LL linerless applicator

the option of linking the machine to weigh scales for variable weight products.

The Nobac 400v sleever is an applicator focused entirely on C-wrap applications and is capable of labeling irregular packages such as whole birds. It maintains close control of both ends of the label in the feed system, improving positional accuracy of the label around the package. The feed system is independent of label length and conveyor speed, allowing longer labels to be run without loss of throughput.

Thermal transfer printers can be optionally fitted. These print at the point of application, minimizing wasteful printed labels on changeovers. The machine can also be interfaced with a weighing machine to print variable data if required.

RITRAMA

In 2013 Ritrama launched Core Linerless Solutions (CLS) with the aim of revolutionizing the use of linerless technology. The system was developed by Ritrama in collaboration with labeling equipment supplier ILTI and is targeted at high volume global end users in the home, personal care and beverage industries. Ritrama regards CLS as an evolution of pressure sensitive technology rather than a replacement.

Indeed, the supply chain for the Core system is exactly the same as conventional pressure-sensitive labels in terms of efficient wide web silicone and adhesive coating and laminating of liners and face films by Ritrama.

Label converters simply print a laminate, without die-cutting, at full press speed, and this is converted on the specially designed RG20 converting machine module (Figure 7.7), which encapsulates the print under the turned around liner, leaving the silicone coating on the face. The self-wound rolls are then micro-perforated.

The Proper-LL linerless applicator built by ILTI (Figures 7.6 and 7.8) allows easy changeover between

Figure 7.7 - The purpose-built RG20 converting module that transforms standard self-adhesive material into a single linerless label web

Figure 7.8 - ILTI linerless applicator on a production line

conventional pressure-sensitive and linerless modules: a trolley-based plug and play unit can be exchanged in under a minute.

In trials to date, line speeds up to 500 bottles/minute have been achieved, although ILTI says that double this speed is possible upon request. Multiple label types can be applied on the same container.

OTHER LINERLESS TECHNOLOGY

Although this book is solely concerned with current pressure-sensitive linerless solutions, it is worth noting a number of other linerless technologies introduced in the recent past. Introduced by companies such as NuLabel and Polykote, these use heat-activated adhesive materials. Avery Dennison has an adhesive that remains pressure-sensitive after application.

Chapter 8

Label application troubleshooting tips and guidelines

While millions of pressure-sensitive labels are satisfactorily applied to all kinds of products and shapes each year there are nevertheless occasions when things go wrong: they may not dispense properly from the liner, or start to dispense in the wrong place. There may be a web break, labels may not adhere properly to a surface or shape, the labels may wrinkle or form ridges, or a label might tear or bulge. Other problems might include the label peeling, curling or lifting at the edge; even falling off.

In general, when application problems do occur these may be related to one or more of the following:

1. Label customer and label usage point of application issues, including label handling, reel change downtime and storage
2. The nature and type of pressure-sensitive material (face, adhesive or liner) being used
3. The label converting process, particularly die-cutting, missing labels, edge nicks, matrix waste not removed, incorrect rewinding tension, handling, storage and shipping
4. Problems on the applicator.

Information provided by Catchpoint indicates that almost half of pressure-sensitive quality faults resulting in packaging line downtime are related to the laminate converting process, particularly with Glassine or Kraft liners. Percentage figures for these quality faults are:

• Die strike	12.2%
• Missing labels	7.8%
• Silicone voids in the liner	5.2%
• Edge nicks	4.8%
• Incorrect winding tension	4.4%
• Bursts through substrate	3.9%
• Label stuck onto back of liner	3.5%
• Matrix waste not removed	3.5%
• Inclusions in the liner	2.6%
Total	**47.9%**

The percentages above will obviously vary between different label application and packaging production lines, as well as over time. The critical end user question is what effect do these and other issues have on labeling production and on label line downtime?

We now examine each of these problem areas and their effects on the label applicator system.

LABEL CUSTOMER AND LABEL USAGE CONSIDERATIONS

To minimize potential application problems, it is important that the label converter obtains in advance as much relevant information from the label-user customer as possible. In particular the converter will ideally need to know:

- The type of container or product that the label is to be adhered to – glass bottle, plastic bottle, paperboard carton, metal surface, tube, flexible pack, etc. Is it rigid, flexible, squeezable? Any of these can have an impact on the label face material or adhesive to be used
- Is the surface being labeled rough or smooth? Rough surfaces may need a thicker or more aggressive adhesive on the label
- The size and shape of the container – small, large, round, square or rectangular, contoured, indented, curved, etc. Again, important in relation to the label adhesive or face material being used
- If the product to be labeled is curved, what is the curvature of the surface – the greater the curvature, the higher the adhesive tack needed to hold the label in position
- If the surface being labeled is plastic, what type of plastic? Are there any plasticizers likely to be present? Some adhesives and plastics 'disagree' with each other and this can deteriorate certain adhesives and cause labels to lift or fall off
- Is the label to be permanent or removable? Different adhesive requirements
- Where will the labels be stored and for how long before they are applied? What are the storage conditions?
- Are the labels being hand or automatically applied? Containers that have been handled with bare hands might lead to labels failing
- Is the surface to be labeled clean, dry (or wet) and free of debris, oil, cleaning or processing solvents or chemicals? Surface contaminations may lead to adhesion problems
- What are the conditions at the point of application? Is the application environment cold, frosty, humid, damp, variable – which can mean that the label adhesive will not properly adhere to the container or surface being labeled
- Are containers, labels and container contents all roughly at the same temperature at the point of application? If not, this again may give an inconsistent label performance
- What happens to the labeled container or product immediately after the labels have been applied? Are they refrigerated? Are they filled after application? How are they handled and stored? What is the product distribution process? All of these may have an impact on label performance.

PRESSURE-SENSITIVE MATERIAL SPECIFICATION

Pressure-sensitive labels, as outlined earlier in this book, are a sandwich construction consisting of different layers:

- The label face material itself, which may be paper, plastic, foil, or other material
- The pressure-sensitive adhesive layer
- A silicone release coating on a backing paper or film that is used to prevent the adhesive sticking to itself or the face material in the reel and during handling.

In general, the application of pressure-sensitive labels tends to be a simpler operation than that for wet-glue applied labels. Apart from accurate positioning, most of the problems that arise with pressure-sensitive labels are due to factors such as the incorrect matching of the label material to the specific end-use requirement, defective converting operations or bad storage, rather than the label applicator or label application process.

For this reason, some of the more important properties of the adhesives and face materials, as well as the conversion operation, are outlined in the following pages.

Pressure-sensitive adhesives. Adhesives used for the production of pressure-sensitive labels are available in a wide range of types that can be used to label almost any kind of product or surface – dry, wet, frozen, glass, plastic, paper, rough, smooth, absorbent, rigid, flexible, stretchable or squeezable, and in different conditions.

Although there are general purpose adhesives that can be satisfactorily used to label a wide range of products and in various application requirements, it is important that the label converter is able to match the adhesive to the specific end-use application if labeling problems, either during, or post application, are to be minimized or avoided altogether.

That is why some or all of the questions set-out in the 'Label customer and label usage considerations' section on the previous page are discussed and agreed between the converter and the user in advance of proceeding with the order.

Pressure-sensitive face materials. A wide range of papers, plastics, metallic foils, metallized and other face materials are used for the production of pressure-sensitive labels. Face material surfaces may be coated or uncoated and, during production may also be printed, varnished, over-laminated, embossed, etc. With the enormous range of face materials, as well as the printing and converting stages, the selection of face material will be based on a compromise between chemical and physical properties, economic considerations and manufacturing possibilities.

It should be recognized that the individual components of a label are not independent in their action. A change in, for example, the thickness or stiffness of the label stock can produce large variations in the properties of the pressure-sensitive layer. It is as well to bear in mind that properties such as peel adhesion or tack are properties of the whole label and not of the adhesion alone.

Factors important to successful dispensing and application of label face materials include the stiffness of the material, or its thinness and flexibility. These will be discussed later in terms of problems that may occur on the applicator itself.

Pressure-sensitive silicone coating. A problem that may be found with siliconized liners is that of silicone voids – small spots where there is no silicone, which in turn allows the adhesive to bleed into the liner fibres. When this occurs it is a laminator issue

which is impossible to detect during label printing and converting. The cause is usually dust during the silicone coating process.

Silicone voids can create a diagonal tear in in the liner at the dispensing beak where the liner is being stressed, causing the applicator to stop. Voids will not break a PET liner, only possibly have an impact on labeling accuracy.

PROBLEMS THAT MAY ARISE FROM THE LABEL CONVERTING PROCESS

Label stock is usually supplied to the converter in the form of reels of pressure-sensitive laminate which will then be printed and die-cut to shape and size in one in-line operation. The printing and the ink drying/curing operations are done on roll-label presses, using standard flexographic, offset, letterpress, screen processes – or combinations of these processes – as well as the latest electrophotographic or inkjet technologies.

Some of the most common converting problems that may arise and impact on the labeling line and label application include:

Die Strike. The most critical operation in terms of successful label application is die-cutting, in which the die must cut cleanly through the label face material and pierce the adhesive layer, but must on no account be clearly marked in the release coating and the liner – particularly with a Glassine or Kraft web – and should certainly not cut into or through the backing release liner. Die-cutting is an operation that requires fine control which needs the laminate and, in particular, the release liner backing material to be of a constant thickness.

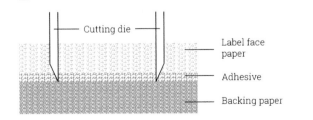

Figure 8.1 - Die-cutting through the label face material and adhesive but not the backing paper

The cutting die has to be tooled to the thickness of the backing liner and, if this varies and the die-cuts into the liner, as in Figure 8. 2 (known as die strike), then the die will have to be re-set. If the die-cutting is too heavy, or too light, and the adhesive has not been cut through cleanly, then labels will not dispense correctly and may not even separate at the beak or stripper plate, instead remaining on the backing liner and carrying on. Thickness variations from place to place in the backing liner will also cause problems.

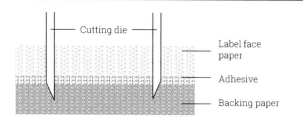

Figure 8.2 - Die-cutting into the backing liner

If the adhesive has not been cut through cleanly or has bled, or there are blobs on the surface, then further problems may arise due to the applicator rollers gumming up.

Web breaks on the applicator may have a range of causes: cuts in the backing liner, bad handling, tears in the edges of the backing, the unwind and reel-up being misaligned, or the clutch on the unwind not working properly or being badly adjusted.

If the label roll has been reeled too tightly or, again, the die-cutting has been carried out badly, then labels can move, becoming displaced on the backing liner. Mislocation will also be caused by inaccurate conversion, incorrect spacing, bad slitting and an inconsistent web width.

Missing labels in the web, with or without evidence of location, may again be due to poor quality die-cutting or excessive adhesive bleed due to either high coating weight in the original laminate or excess pressure in the printing process. The label may have been carried into the matrix waste and the gap not picked-up during inspection.

Some applicators are able to compensate for missing labels, while on others there will be no label,

or perhaps cases of inaccurate label placement. There may sometimes be a double label.

Reel and liner edge nicks. These will show as a tear in the edge of the liner and will invariably snap the liner at the application beak. The problem is caused by damaged or poor quality slitting knives affecting the edge of the liner, or by damage to the reels during packing and handling on delivery to the line.

Although reels with PET liners are more durable, it is important that slitting knife quality is maintained.

Incorrect winding tension. If tension is incorrectly set in the printing press or in the slitter rewinding process then it is possible that reels may telescope, ridges form, or that reels may slip and feed intermittently. Ridging may be particularly evident in plastic materials. In the application process, ridged labels will cause incorrect dispensing due to the optical sensor reflecting on the ridged labels.

Low tension may cause accuracy issues, while winding too tight, particularly with a high adhesive coating weight, may cause adhesive bleed, affecting applicator path rollers.

Labels on the back of the liner. This can be due to a winding tension that is too high, combined with a high adhesive coat weight, which in turn causes adhesive bleed in the reel and a sticky residue to form around the label edges, leading to labels being plucked off the release face and sticking to the back side of the liner. The labels are then not dispensed and there will be a build up of adhesive on the applicator path roller.

Reels with PET liner are susceptible to high tension and adhesive bleed.

Matrix waste in the label reel. Poor control of the waste matrix by the operator - or worn tooling - leads to material filling, or partially filling, the gap between labels. In the applicator the label sensor may align to this matrix and deliver two labels. Most applicators will stop at a default length, but this in not exact to the label size. Edge of reel waste may end up sticking to the product being labeled. It is most likely to occur with paper labels.

Dirt inclusions in the liner. Seen as dark spots in the liner and therefore not to be confused with silicone voids. May lead to tears at the applicator beak as die strike is likely against the inclusion. Not seen with PET liners.

PROBLEMS ON THE APPLICATOR

Obviously applicator machines and labeling lines must be set up correctly if the label application process is to proceed successfully. But even if this has been done, problems can still arise as a result of defective converting operations, bad storage or the incorrect choice of materials to match the end-user requirements.

To summarize: the main problems that can arise at the applicator are due to dispensing too early or too late, and mis-location or breaks in the backing liner. Providing the applicator has been set up correctly, most problems can be traced to label stock that has not been matched to the equipment through to faulty label conversion or issues of storage. Some particular issues to note are as follows:

Labels detaching too early. The right place for labels to be peeled from the backing liner is at the beak or stripper plate. During its passage through the applicator, the web is taken through small diameter rollers. If the label face paper is very stiff or the release value is too low, the labels may shoot straight on, detaching themselves from the backing too early rather than bending around the roller with it. This can be seen in Figure 8.3.

stiff labels: the only adjustment that can be made on the applicator is the angle of the stripper mechanism.

Surface contamination. Once the label has peeled off it must stick firmly to the container or product surface under conditions of light pressure for a very short period of time. Although adhesives are extremely versatile with regard to the range of materials to which they will bond firmly, contamination of the container or product surface with wax, oil, grease, silicone or other release agents, or moisture, will affect adhesion. For satisfactory adhesion the surface should be dry and free from contamination. Clean bottles are imperative for successful label application.

If surfaces to be labeled are wet, frosted, cold or hot, then special adhesives suited to the particular application surface or application environment will need to be used.

Container curvature. A further point to consider with regard to containers is the curvature of the surface – the greater the curvature, the higher the adhesive tack needed to hold the label in position. The problem will be worse with stiff label materials which may start to lift away from the labeled surface (Figure 8.4).

Figure 8.4 - Stiff label materials on small radius containers may start to lift at the edges

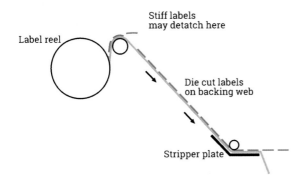

Figure 8.3 - Stiff labels may detach too early

In effect, the small roller is acting as a stripper plate. A very fine balance of properties is needed for

A flexible label material will conform more easily to the container curvature, reducing the tendency of the label to lift or 'butterfly'. Butterflying after application is due to the face material being too stiff, incorrect adhesive, surface contamination, wrong label shape for the container or insufficient roller, brush or tamp pressure.

Labels wrinkling. If applied labels contain folds or ridges, these indicate that wrinkling is occurring. The main cause of wrinkling is the use of the wrong label material for the container. The label converter needs to know whether the container being labeled is plastic or glass, and whether it is flexible or rigid. Irregularities in a container or improper application may also cause wrinkling.

When tubes are being labeled, ridges or folds may again occur. This is commonly caused when a semi-rigid label material (such as polypropylene) has been used, rather than a material that is fully squeezable and flexible in all directions (such as polyethylene or polyolefin).

Labels peeling or curling. This can occur if labels are applied in cold or damp environments or if the application surface is damp, dirty or greasy. Peeling or curling may also occur if the label adhesive has had insufficient time to 'set' before a labeled product is refrigerated (which may typically be 24 hours).

Labels flagging. If a bulge occurs along the edge of a label once it has been applied, this is sometimes called 'flagging'. It can occur if the label is too large for a container or if the label is not the right shape to fit a curved or irregular-shaped container. Labels must be the right size and shape for a container; something which should be determined in advance with the label converter.

REMEDIAL ACTION WHEN LABELING PROBLEMS OCCUR

Despite the care taken during label printing and converting, and in the setting-up of the applicator line, it is still possible that problems may arise. When these do occur it may be possible to take corrective action. These can be summarized as follows:

1. Check the applicator, ensure all settings are correct, and make any adjustments required
2. Examine the label reels. If they are defective this may be apparent on inspection. If a defect is found it does not necessarily follow that all reels are affected, so it will be worthwhile examining and isolating any affected ones
3. Check that the storage conditions are right and that the reels have not been stored for too long or on edge
4. Ensure that containers or other products to be labeled are clean and dry – probably more important than most users realize. If not, see whether it is possible to install hot air blowers, or bring the container or product being labeled into the labeling hall to be conditioned in good time before application takes place.

To avoid unnecessary problems it is advisable that the label purchaser/label user has some knowledge of the basic types of labels available and their properties, and that there is liaison throughout from the label materials supplier, the label converter to the label purchaser and final label user.

Such an exchange of information can be invaluable for ensuring trouble-free labeling. Although this is perhaps stating the obvious, it is all the more amazing that these common-sense practices are not followed more often.

Index

Made in the USA
Monee, IL
27 October 2021

80837096R00043